AN ANNOTATED BIBLIOGRAPHY ON RURAL TRANSPORT

Edited by Niklas Sieber

IFRTD
The International Forum for Rural Transport and Development

Practical Action Publishing Ltd
25 Albert Street, Rugby, CV21 2SD, Warwickshire, UK
www.practicalactionpublishing.com

© International Forum for Rural Transport and Development, 1997

First published 1997\Digitised 2008

ISBN 13 Paperback: 9781853394188
ISBN Library Ebook: 9781780444505
Book DOI: https://doi.org/10.3362/9781780444505

All rights reserved. No part of this publication may be reprinted or reproduced or utilized in any form or by any electronic, mechanical, or other means, now known or hereafter invented, including photocopying and recording, or in any information storage or retrieval system, without the written permission of the publishers.

A catalogue record for this book is available from the British Library.

The authors, contributors and/or editors have asserted their rights under the Copyright Designs and Patents Act 1988 to be identified as authors of their respective contributions.

Since 1974, Practical Action Publishing has published and disseminated books and information in support of international development work throughout the world. Practical Action Publishing is a trading name of Practical Action Publishing Ltd (Company Reg. No. 01159018), the wholly owned publishing company of Practical Action. Practical Action Publishing trades only in support of its parent charity objectives and any profits are covenanted back to Practical Action (Charity Reg. No. 247257, Group VAT Registration No. 880 9924 76).

Design and composition: Krishan Jayatunge

Reasonable efforts have been made to publish reliable data and information, but the author and publisher cannot assume responsibility for the validity of all materials or for the consequences of their use.

The manufacturer's authorised representative in the EU for product safety is Lightning Source France, 1 Av. Johannes Gutenberg, 78310 Maurepas, France.
compliance@lightningsource.fr

CONTENTS

ACRONYMS AND ABBREVIATIONS	6
FOREWORD	8
1. **RURAL TRANSPORT INFRASTRUCTURE**	10
2. **INTERMEDIATE MEANS OF TRANSPORT**	20
3. **ANIMAL POWER**	26
4. **RURAL TRANSPORT SERVICES**	40
5. **RURAL TRANSPORT PLANNING**	52
6. **FINANCIAL AND INSTITUTIONAL ISSUES**	60
7. **TRANSPORT AND DEVELOPMENT**	70
8. **GENDER AND TRANSPORT**	84
9. **MISCELLANEA**	96
10. **PROMOTIONAL MATERIAL**	102
11. **HOW TO GET HOLD OF THE DOCUMENTS**	106
INDEX BY AUTHOR	110
INDEX BY TITLE	114
INDEX BY KEYWORD	123

ACKNOWLEDGEMENTS

This annotated bibliography was initiated by the International Forum for Rural Transport and Development and the ILO as an 'expert selection' of relevant documents in a number of key areas. It is a combined effort of several members of the Forum network who contributed by identifying key documents in their area of expertise and annotating them. We acknowledge with thanks the contributions of Professor Paul Starkey, who did the annotations for the animal power section, Mrs. Jo Leyland (née Doran) for gender and transport, Mr. Ron Dennis for intermediate means of transport and Dr. Simon Ellis for rural transport services. The following provided us with lists of documents to be included in the bibliography: Ms Fatemah Ali Nejadfard, Ms Christina Malmberg Calvo and Mr Gary Taylor. These were annotated by Niklas Sieber and Priyanthi Fernando at the Secretariat. Ms Ana Bravo collated the different sections into one document and compiled the indexes. The design and layout was done by Krishan Jayatunge from Sri Lanka.

ACRONYMS AND ABBREVIATIONS

AGAP	Animal Production Service
AMRU	National Association for the Development of Rural Women
ATNESA	Animal Traction Network of Eastern and Southern Africa
BBLL	Bridge Building at the Local Level
CTA	Technical Centre for Agricultural and Rural Co-operation, Wageningen
DBSA	Development Bank of Southern Africa
DTU	Development Technology Unit, University of Warwick
EASTS	International Conference of the Eastern Asia Society of Transportation Studies
FAO	Food and Agriculture Organization
GATE	German Appropriate Technology Exchange
GIS	Geographical Information System
GNP	Gross National Product
GPRTU	Ghanian Private Road Transport Union
GTZ	German Agency for Technical Cooperation
ILO	International Labour Office (Geneva)
IMT	Intermediate Means of Transport
IRAP	Integrated Rural Accessiblity Planning
IRDP	Integrated Rural Development Programme
ITDG	Intermediate Technology Development Group
ITDP	Institute for Transportation and Development Policy
ITP	Intermediate Technology Publications
KHARDEP	Kosi Area Rural Development Programme
LTC	Least Total Cost Planning
NGO	Non-Government Organization
NMT	Non Motorized Means of Transport
NTRC	National Transport Research Centre, Pakistan
pkm	Passenger kilometre: passengers transported per kilometre
RMI	Road Maintenance Initiative (Part of SSATP, World Bank)
RRA	Rapid Rural Appraisal

RTTP	Rural Travel and Transport Program (Part of SSATP, World Bank)
SANAT	South African Network of Animal Traction
SKAT	Swiss Centre for Development Co-operation in Technology and Management
SSAfrica	Sub-Saharan Africa
SSATP	Sub-Saharan Africa Transport Program (World Bank)
tkm	Tonne kilometre: tonnes transported per kilometre
TOC	Truck Operating Costs
TRL	Transport Research Laboratory
TRRL	Transport and Road Research Laboratory (TRL's previous title)
UNCHS	United Nations Centre for Human Settlements
UNIFEM	United Nations Development Fund for Women
USAID	US Agency for International Development
VLTTS	Village Level Travel and Transport Surveys (Part of RTTP, World Bank)
VOC	Vehicle Operating Costs

FOREWORD

The vast majority of rural communities in developing countries continue to experience forms of disadvantage, which commonly include isolation. Remoteness, lack of contacts and information and lack of access to basic goods, social services and economic support prevent such communities from improving their living conditions.

Rural transport policies, planning systems and investment programmes can play a crucial role in reducing isolation, thus the wellbeing of rural people. The enhancing potential for contributing to economic and social development through measures in the transport sector is large indeed if it is considered that transport accounts for a big share of overall investment in most developing countries. There is, however, a need to change the way rural transport interventions are planned and implemented. It is a fact that conventional models based on public provision and maintenance of road networks and private sector supply of motor vehicles have seldom resulted in benefits for the poor because they have been justified on purely economic criteria. An emerging view is that rural transport has to be seen in the broadest sense as providing access to address the actual needs of rural populations.

The debate is open and can refer to the substantial amount of work which has been carried out over the last ten years with the involvement of several organizations, institutions and individuals and support from a number of donors. As a result of this and previous work on rural transport and accessibility, substantial and substantive documentation is now available, though it is scattered in different locations which are often unknown to potential users. Therefore, in order to make existing information easily accessible to all those who might be interested, the International Labour Organization (ILO) and the International Forum for Rural Transport and Development (IFRTD) have jointly engaged in the production of this annotated bibliography.

The present book brings together the abstracts of a selection of relevant documents on a number of transport-related areas, namely infrastructure, intermediate means of transport, transport services, policy and institutional issues, financial and planning issues and gender issues. Its principal purpose is to reach a wider audience of rural development and transport planners and practitioners, academic and training institutions, and concerned government agencies and NGOs in order to broaden the debate and influence thinking on rural transport policies which are of direct benefit to the rural poor.

> Development Policies Department
> International Labour Office, Geneva
> April 1997

1. RURAL TRANSPORT INFRASTRUCTURE

DIXON-FYLE, KANYHAMA AND IRENE FRIELING (1990)
Paths in rural transport: A study of Makete, Tanzania
International Labour Office, World Employment Programme, Geneva; 74 p.
Available from: ILO/Geneva

Keywords:
low-cost infrastructure,
paths,
technology,
Tanzania,
household survey,
traffic survey

This report documents the findings of a survey of footpaths in the Makete District in Tanzania undertaken by the authors for the Makete Integrated Rural Transport Project in October 1988 (see Barwell and Malmberg-Calvo 1988; Sieber 1996). The path survey was the first of its kind in the context of a transport project. The report is intended to be the basis on which future interventions on paths can be based. It also provides information and insights on rural transport issues.

An initial quantitative definition of the transport demand and transport patterns indicated that most of rural transport takes place on the paths in and around villages. The survey on which this report is based addresses questions of the problems of access related to the condition of the path network, the interest of footpath users to improve the paths by self-help, the possible technical solutions and the potential impact of improvements in terms of time and effort saving for users. There is a chapter that describes the problems and provides options for their solution. Another chapter synthesizes the information from the different village studies. The survey indicates that a large part of the path network is in poor condition. Water, firewood sources, field and grinding mills are often only accessible with great difficulty, especially during the wet season. Since the overall objective of the Makete Integrated Rural Transport Project is to find sustainable methods to reduce the time and effort people in the district spend on transport, the report concludes that improving paths is contributing to that objective. The report makes recommendations of improvements that would create year round, safe and easy access for people on the major path routes in and around the villages.

The report includes two detailed annexes. The first one describes the technical details of the suggestions proposed for path improvement. Annex 2 provides detailed village studies that provide a brief description of the physical location of the village, the agricultural and economic characteristics, the transport characteristics, proposals for action. A last section in each village study describes the attitude of the villagers to transport provision and to project interventions.

EDMONDS, GEOFF AND JAN DE VEEN (1992)
'A labour-based approach to roads and rural transport in developing countries'
International Labour Review Vol. 131, No. 1: 95-110,
Special Issue, Geneva.
Available from: ILO/ASIST, Nairobi

The article gives an introduction to the rationale and organization of labour-based road construction and maintenance works. Firstly the authors describe the various factors influencing the technology choice (e.g. wage levels, availability of labour, etc.) and draw a decision tree using these. The next chapter discusses the use of local resources. For the involvement of the private sector, appropriate contractual procedures have to be developed and an adequate cash flow must be secured. Before labour-based works can be conducted, appropriate light equipment has to be developed. Different organizational arrangements are discussed for road maintenance: individual or collective arrangements, use of petty or small-scale private contractors and agreements between government and communities.

Keywords:
**technology,
roads,
maintenance,
institutions**

ENGLER, MARKUS (1994)
Participation in rural infrastructure programmes: A process oriented approach to bridge building at the local level in Nepal.
Landwirtschaftliche Beratungszentrale, Lindau, Switzerland; 28 p.
Available from: ILO/ASIST, Nairobi

Keywords:
community participation, capacity building, local knowledge, bridges, planning, low-cost infrastructure

The paper describes the Bridge Building at the Local Level (BBLL) programme in Nepal. The programme was a piloted by Helvetas and the Swiss Development Co-operation to test a new concept for supporting the construction of trail bridges. Initially regarded with some scepticism by the Nepalese government, the value of the programme's approach is now (unofficially) accepted. The physical output of the programme is now comparable with the bigger government programmes.

The BBLL concept is based on the idea that the beneficiaries of a programme must be given the opportunity to define the problem themselves and to find solutions appropriate to their circumstances. The BBLL programme considers itself a catalyst for triggering processes of co-operation within a network of institutions. The programme offers flexible but clearly defined support, interlinks existing organizations and enhances the efficiency of local organizations.

The paper begins by outlining the basic approach and philosophy of the BBLL programme and its significance within the context of development policy in Nepal. It goes on to describe in detail how the approach and philosophy is translated into action. The description includes details of the various steps in the implementation process, the institutional arrangements, the process of co-operation and the package of support (called the 'toolbox') provided by the programme. The text is supplemented with real life experiences from the programme and some interesting line drawings. The author continually emphasizes how the programme has managed to maintain the importance of local knowledge and skills, local organization and local ownership of the technology.

The paper concludes with some valuable lessons for other programmes interested in implementing rural infrastructure through community participation. It provides an interesting case study for everyone interested in community participation and also some key pointers as to how participation can be achieved.

HINDSON, JACK: REVISED BY JOHN HOWE AND GORDON HATHWAY (1983)
Earth roads: A practical guide to earth road construction and maintenance
IT Publications, London; 123 p.
ISBN 0 903031 84 1

The book fills the need for a simplified description of road construction and planning at the most elementary level. The author attempts to explain how earth roads can be constructed and maintained in a way that would prevent their progressive and often rapid destruction by erosion due to the uncontrolled flow of water. The emphasis throughout the book is on the proper control of water reaching the road, and on the basic principles of soil conservation. The focus is quite deliberately on roads designed to carry a few dozen vehicles a day at most. Fully engineered gravel or bitumen surfaced roads are not considered.

Keywords:
earth roads, technology, Zambia, low-cost infrastructure, planning, maintenance

The text covers two different basic designs of earth roads called, for convenience, 'village' and 'market' roads. These are not precise definitions. The essential difference is the level of usage, from a few vehicles per day or week (village) to perhaps fifty vehicles per day (market).

The methods described in the book are based on the accumulated experience of building and maintaining earth roads for over twenty years in the northernmost parts of Zambia. The methods emphasize the achievement of low construction costs by adopting low-speed natural alignment for the road, using local, labour intensive technology where possible and minimizing the cost of earth moving. The author recognizes that in addition to conventional motor vehicles, such roads may be used by a range of much simpler vehicles including wheelbarrows and handcarts, animal drawn carts and bicycles.

The editors of the original publication have updated some of the figures, standardized some of the terms and shortened the text so that the publication could be more affordable to readers. The text is divided into four major parts. The first part deals with the drainage principles and techniques. Part two covers the planning of the road. The third section describes construction methods and the final part deals with subsequent maintenance. There is one appendix that describes the operation and use of some simple pieces of surveying equipment and one that provides a glossary of road construction terms.

KRÄHENBÜHL, J. ET AL. (1983, 1990, 1992)
Survey, design and construction of trail suspension bridges for remote areas
SKAT/HELVETAS, St. Gallen, Switzerland
Available from: SKAT

Keywords:
bridges,
low-cost
infrastructure,
technology,
costs,
traffic survey,
manufacture

Long-standing experience in the construction of pedestrian bridges in the Nepalese hills by the Suspension Bridge Division of His Majesty's Government of Nepal, by the Swiss Association for Technical Assistance in Nepal, and by Helvetas, Swiss Association for Technical Assistance, Zurich, Switzerland, forms the basis of this manual.

The manual is published in five parts, containing all the necessary technical, legal, organizational, and economical information for survey, design and construction of pedestrian bridges. The five volumes are:

GROB, A., J. KRÄHENBÜHL, A. WAGNER (1992)
  Volume A: Design
  ISBN 3 908001 36 6; 375p.
KRÄHENBÜHL, J. AND A. WAGNER (1983)
  Volume B: Survey
  ISBN 3 908001 63 3; 325p.
BASNET, C., J. KRÄHENBÜHL AND P. BHATTA
  Volume C: Standard Design Drawings
  (Volume not currently available until revision is completed)
PANCIOTTO, D. (1983)
  Volume D: Execution of Construction Works
  ISBN 3 908001 64 1; 287p.

KRÄHENBÜHL, J. AND H.P. MAAG (1983)
Volume E: Costing and Contracting
ISBN 3 908001 65 X; 289p.

The volumes include:

- bridge site selection criteria and feasibility surveys
- geological survey and soil investigations with special emphasis on geologically unstable areas
- detailed procedures for bridge planning, design and structural analysis including a selection of related computer software
- a set of 164 standard design drawings
- descriptions of the execution of construction works, erection of superstructures, the necessary instruments, tools and machinery
- check lists of acceptance for all parts of the works
- data and standard forms for cost estimating and rate analysis.

Of all the different possibilities of river crossings, the unstiffened cable bridge has proved to be the most feasible and economic bridge type in the topographically, geologically and hydrologically difficult areas in the foothills of the Himalayas. Two bridge types have been selected for standardization: a suspension bridge with towers and gravity anchorage called "Suspension Bridge" and a cable bridge without towers where the walkway is fixed directly on top of the load-bearing cable called "Suspended Bridge".

The technology has been selected on an intermediate level appropriate for adaptation in developing countries. All the components of the bridges have been designed in such a way that they can be manufactured in steel workshops with only basic installations, transported by manpower and erected only with the help of simple pulling and lifting equipment.

MARSH, D.K.V AND GARY TAYLOR (1982)
The movement of goods and people in the Kosi Hills
KHARDEP report No. 37 Kosi Hill Area Rural Development Programme, Nepal; 56 p.
Available from: ILO/ASIST, Nairobi

Keywords:
low-cost infrastructure, planning, traffic survey

The Kosi Area Rural Development Programme (KHARDEP) was established to assist in the development of the four hill districts of the Kosi Zone in eastern Nepal. One of the major constraints to the development of these districts is the lack of roads and the difficulty of communication. However, there is a large volume of traffic on the main foot trails. This report is a synthesis of the findings of a series of traffic surveys conducted in 1981 with the objective of assisting KHARDEP funds in the communication sector. The broad aims of the surveys were to identify main routes, quantify traffic volumes and seasonal fluctuations, investigate migration and provide baseline data for use in future studies prior to the opening of a major roadway. The study is unique in Nepal and rare in transport literature in general. While the information is specific to Nepal, the methodology of the survey and the analysis could provide important guidelines for similar work to be carried out elsewhere.

SELVARASA, K. (1992)
Manual on the improvement/maintenance of footpaths. Makete integrated rural transport project.
(first draft) 20 p.
Available from: ILO/ASIST, Nairobi

Keywords:
paths, maintenance, drainage, water crossings, low-cost infrastructure

The manual presents simple and understandable guidelines on the methods of improvements and maintenance of paths using locally available resources. It is based on the conditions of the path network in and around villages in Makete District in Tanzania and the problems caused by the district's topography, soil conditions and wet climate, i.e. steep slopes, slippery soils, river and marshy crossings.

The author describes two types of problems: major problems of drainage, steepness and water crossings and the minor problem of invading vegetation. Solutions are provided for these problems. The manual describes techniques for surface drainage, sub-surface drainage, catchwater drains, anti-slip surface and side erosion control; it provides techniques for relocation, for the construction of steps, foot bridges and access over marshy areas. The text is supplemented with diagrams and construction details.

The author also provides guidelines for maintenance of the paths from the perspective of routine maintenance that must be done on a continuous basis and periodic maintenance at intervals of several years.

TAYLOR, GARY (1994/95)
'Improving paths and tracks'
Part 1: *Appropriate Technology* Vol. 21 No. 1, June 1994: 17-20, AT Brief No. 8.
Part 2: *Appropriate Technology* Vol. 22 No. 1, August 1995: 17-20, AT Brief No. 12.

The technical brief provides a clear, concise and comprehensive overview of how improvement to paths and tracks can be made. The first part describes the identification of problems on paths and tracks, the items to consider in the planning of path and track improvements, recommended standards to adopt; methods of constructing a path or track including the surfacing materials and the organization of the work. It stresses the importance of user participation. In the seven charts and diagrams available with the text, the author provides technical guidelines for deciding several aspects of path construction and maintenance, including typical labour productivity.

The second part focuses on paths and tracks that cross wet and marshy areas. With the aid of illustrations, the brief provides technical information on three techniques by which paths across wet or marshy areas can be improved: stepping stones, rafts or boardwalks and 'turnpike section'. The brief also includes a short section on improving paths in very sandy areas.

Keywords:
paths, design standards, organisation of work, marshy areas, technology, low-cost infrastructure

US DEPARTMENT OF AGRICULTURE (1995)
Forest service handbook
Amendment No. 11
Available from: ILO/ASIST, Nairobi

Keywords:
**trails,
construction,
maintenance,
institutions,
low-cost
infrastructure**

This handbook looks at the administration, trail construction plans, construction, maintenance and operation procedures for the forest development trail system in the United States.

In the US, trails are the means of access to more remote areas where the construction of roads cannot be administratively or economically justified, or funds for roads are not available. Trails are also used where the type of travel enhances public enjoyment of the forest environment and utilization of fish and wild life resources.

The chapter on administration details all the policy, procedures and criteria for administering the system. The chapter on trail construction plans looks at different types of surveys and the design of trail systems. The chapters on construction and maintenance goes into the technical details of clearing, tread construction, drainage, surfacing, structures, safety and reporting. The chapter on maintenance also includes a section on organization, work plans and condition surveys.

The authors of the handbook continuously highlight the differences between roads and trails, particularly in terms of the choice of options available. The handbook is based on the US system but it provides good technical information supplemented with clear diagrams where necessary. It is a useful resource fo those wanting to adapt trail technologies for other contexts.

THE WORLD BANK (1976)
Haulage using aerial ropeways: World Bank study on the substitution of labour and equipment in civil construction
Technical Memorandum No. 22, two appendices, Washington D.C.; 39 p.

Available from: ILO/ASIST, Nairobi

Although this paper is now quite old, it gives a sound technical introduction on the construction and use of ropeway systems. It is based on fieldwork carried out during studies in India and Indonesia with the scope of using aerial ropeways in labour-intensive construction work. It is shown that in appropriate circumstances a simple ropeway installation can give significant reduction in the unit cost of haulage compared to manual load carrying.

Keywords: **ropeways, technology, efficiency, infrastructure, Indonesia, India**

The document gives detailed technical descriptions about the design, construction and use of ropeways. The ropeways tested were used for haulage of sand, stones and soil using manpower or gravity in different terrain. Various technical aspects of the ropeways are compared to load carrying with sacks or headbaskets. The ropeways were roughly halving the labour input. A comparison of total costs was not possible.

The appendices give technical details on the tested ropeways.

2. INTERMEDIATE MEANS OF TRANSPORT

DENNIS, RON AND ALAN SMITH (1995)
Low-cost load carrying devices: The design and manufacture of some basic means of transport.
IT Publications, London; 180p.
ISBN 1 85339 265 0

Keywords:
**IMT,
NMT,
design,
manufacture,
technology,
low-cost
vehicles**

This book describes and discusses the technical characteristics of a range of basic transport technologies which are at the low-cost end of Intermediate Means of Transport. Chapters are included on carrying devices for back and shoulders, wheelbarrows and handcarts, carriers and panniers for bicycles, and panniers and sledges for animal-based transport. An initial chapter compares the characteristics and applications of these transport technologies and presents a guide for their selection. The subsequent chapters provide detailed guidelines on the design and manufacture of each of the technologies, mainly in terms of small-scale production by artisans and small workshops. The book contains many illustrations of the technologies in use, and drawings showing clear design details and methods of manufacture. It is therefore of interest to planners and development agencies providing an awareness of the various technologies, and also to those wishing to set up production of the technologies.

HATHWAY, GORDON (1985)
Low-cost vehicles: Options for moving people and goods
IT Publications, London; 106p.
ISBN 0 946688 02 8

This A5-size book is a catalogue of low-cost vehicles (IMT) which are used in various parts of the developing world. The book is in two parts: the first part gives a brief overview of transport needs and patterns of rural transport and of the applications and benefits of low-cost vehicles; the second lists a wide range of low-cost vehicles which are used in rural transport with brief details of each. The vehicles are classified into the following groups: carrying devices; wheelbarrows and handcarts; animal-based transport, panniers, sledges and carts; pedal-driven vehicles, bicycles and tricycles, bicycles with sidecars; motorcycle and motor scooter based vehicles; trailers for bicycles and motorcycles; and basic motorized vehicles such as single-axle tractors, three-wheelers and utility vehicles. Each entry comprises a photograph, a short description and brief specifications. The book is intended to provide an awareness of options and their possible uses –it does not provide in-depth details or design information. It is a valuable first-step reference on IMT options.

Keywords:
MT,
NMT,
low-cost vehicles

IT TRANSPORT LTD. (1996)
Promoting intermediate means of transport
SSATP Working Paper 20, World Bank, Washington D.C.
Available from: World Bank SSATP

Keywords:
**IMT,
credit,
planning,
accessibility**

This paper reviews field experience to derive lessons and develop guidelines on the selection, introduction and dissemination of IMT. The guidelines are presented in the form of a decision-making model which has four stages:

- Contextual factors – these are the range of factors which need to be considered in the selection of appropriate IMT: primary factors such as topography, infrastructure, demography and the local economy; secondary factors such as local culture, community organization, manufacturing resources and financing of production; and non-project factors such as institutional, policy and regulatory issues.
- Access issues – this considers the role of IMT in the wider context of accessibility planning for a region, bringing in other options of infrastructure improvement and non-transport interventions.
- Defining needs and targets – based on an identification of access needs and contextual factors, needs and targets are defined and a programme to achieve these outlined.
- Action planning – this stage develops a detailed plan of action for increasing the use of IMT to improve access.

The guidelines use available experience to provide advice to planners on working through the above stages of a systematic approach to the introduction and/or dissemination of IMT. A separate section is included on establishing credit facilities because of the importance of credit in marketing IMT.

IT Transport Ltd. (1988)
The design and manufacture of low-cost motorized vehicles
IT Publications, London; 190p.
ISBN 1 85339 070 4

This is a reference manual on the design and manufacture of a range of low-cost or intermediate motorized vehicles that could be produced in developing countries. These vehicles are particularly suited to providing low-cost transport services over short to medium distances in rural areas and are important in giving access to markets, rural centres and other facilities external to the village environment. The vehicles covered are: motorcycle attachments, sidecars, trailers and a 4-wheel conversion; small-engined 3-wheelers, motorcycle conversions and small diesel-engined vehicles; small-engined 4-wheelers; and single-axle tractor-trailers. An introductory chapter compares the characteristics and possible applications of the different types and then each classification is described in considerable technical detail in subsequent chapters. Technical details of all important features of each type of vehicle are presented, illustrated by photographs and conceptual sketches, but actual detailed design drawings are not included. Guidelines are given on organization of manufacture and on the design and construction of the main features of low-cost vehicles, such as frames, bodywork, suspension, wheels and axles. A compendium of commercially available vehicles is included, comprising photographs, detailed specifications and supplier details. Overall, this is a comprehensive and very useful technical reference manual on these types of vehicles.

Keywords:
**IMT,
transport services,
manufacture,
technology,
low-cost vehicles,
motorized vehicles**

RIVERSON, JOHN AND STEVE CARAPETIS (1991)
Intermediate means of transport in Sub-Saharan Africa
World Bank Technical Paper No. 161, The World Bank,
Washington D.C.; 27p.
ISBN 0 8213 1951 5

Keywords:
IMT;
SSAfrica,
planning;
transport
services

This review paper is one of a series contracted by the Africa Technical Department and the SSATP of the World Bank to address rural transport issues in Sub-Saharan Africa. The paper briefly reviews available data on rural transport and argues that improvements to rural infrastructure and low-cost transport services are essential requirements for economic and social development of rural areas. An integrated strategy is proposed with IMT providing transport services at village level and linking up with motorized services operating on the highway system and with improvements in infrastructure to match the needs of the modes of transport. Reasons for the limited use of IMT in SSAfrica are discussed and a new approach to rural transport planning put forward to help to expand the use of IMT. It is argued that the increased use of IMT and other transport services will increase the benefits from improvements in rural infrastructure.

SALIFU, M. (1994)
'The cycle trailer in Ghana: A reasonable but inappropriate technology.'
African Technology Forum Vol. 7, No. 3: 37-40,
Cambridge, U. S. A.
Available from: ILO/ASIST Nairobi

Keywords:
IMT,
bicycle-trailers,
dissemination,
gender,
Ghana,
cultural
constraints

This short article discusses the impact of a programme to introduce bicycle trailers into Northern Ghana, an area where bicycles are widely used. It provides useful lessons on why an apparently appropriate form of IMT may not necessarily catch on with local communities to the extent anticipated. The article briefly outlines the reasoning behind the selection of bicycle trailers and the programme to introduce them, and then goes on to discuss some of the factors which have limited the take-up of the trailers. The main constraints are the high cost of the trailers, the fact that many of the loads to be moved are relatively small and compact and can be carried on a bicycle carrier. There is also quite a rigid division of household tasks. Women are the main transporters but very few women own or even ride bicycles. The article concludes that IMT are needed to relieve the burden of headloading but because of local culture they need to be targeted primarily at women. In the case of bicycles and trailers a long-term strategy is needed to break down the barriers to women using bicycles and to improve their access to bicycles. It should be pointed out that although the initial take-up was slow there are reports that a significant number of trailers are now in use.

SHELADIA ASSOCIATES (1992)
Study of rural transport services provided by motorized three wheelers in Gujarat State, India
Report prepared on behalf of IT Transport Ltd. for the World Bank Sub-Saharan Africa Transport Programme; IT Transport Ltd., U.K., ILO, Geneva.
Available from: ILO/POLDEV Geneva and ILO/ASIST Nairobi

Motorized IMT play an important role in providing low-cost transport services in some Asian countries. This report is a detailed case study of one example in Gujarat State on the West coast of India, where about 27 000 of the vehicles were in use at the time of the study. The vehicle comprises a locally-made steel chassis to which are attached the front fork assembly of a motorcycle and the rear axle of a small car. It is powered by a 6.5hp diesel engine and it can transport up to 1 tonne at average speeds of 30 km/hr. The aim of the report is to provide sufficient data to assess the potential for introducing the vehicle in other regions and detailed information is given on design, types and sources of materials and components used, methods of manufacture and the organization and economics of operation of the vehicles. The report is a valuable source of information on both the technology and the operating characteristics of this type of IMT.

Keywords:
**IMT,
motorized vehicles,
low-cost vehicles,
India**

3. ANIMAL POWER

ANDERSON, MARY AND RON DENNIS (1992)
'Improving animal-based transport in Eastern and Southern Africa'
in: *Proceedings of the First Conference of the Animal Traction Network for Eastern and Southern Africa (ATNESA), Zambia, 1992.*
Available from: ATNESA

Keywords:
IMT,
animal power,
SSAfrica,
socio-economic impacts,
technology

This paper summarises the main issues of animal-based transport and the role it could play in alleviating the transport burden of rural households in Eastern and Southern Africa. In particular it identifies ways in which the production and dissemination of animal-drawn carts could be improved. The paper is based on a review of the experience of animal-based transport in both Africa and Asia and covers pack donkeys and animal-drawn sledges and carts.

Part I of the paper provides an overview of animal-based transport in Eastern and Southern Africa. It examines the role of animal-based transport in the region, reviews technology options, discusses animal-drawn cart design and production issues, and considers socio-economic aspects of cart dissemination. Part II looks in more detail at the technical aspects of cart design covering in particular options for wheel and axle design.

BARWELL, IAN AND GORDON HATHWAY (1986)
The design and manufacture of animal-drawn carts
Technical memorandum prepared for ILO/UNCHS (HABITAT). IT Publications, London; 72p.
ISBN 0 946688 52 4

Keywords:
IMT,
animal power,
carts,
harnessing,
technology

This book provides detailed technical information on the design and manufacture of animal-drawn carts and harnessing arrangements. It discusses wheel types, axle types, bearings and harnessing systems in relation to cart function. The book identifies particular design features that are necessary for specific jobs such as water carrying and human transport. A chapter on manufacturing aspects discusses the practicalities and economics of cart production at village level and on a national scale. An appendix discusses the factors that influence the draft force requirements of a cart and ways of modifying them. The book includes pictures, photographs and technical drawings of carts and harnessing systems from around the world. The book is now quite dated (it was prepared in 1986) but is still a useful resource.

Keywords:
IMT,
animal power,
carts,
harnessing,
technology

BETKER, J. AND H. KUTZBACH (1989)
'Influence of design on the draught force characteristics of animal-drawn carts'
in: D. Hoffman, J. Mari and R. J. Petheram (ed), *Draught animals in rural development, Proceedings of an international research symposium held at Cipanas, Indonesia, 3-7 July, 1989.* ACIAR Proceedings Series No. 27. Australian Centre for International Agricultural Research, Canberra, Australia; pp. 258-263.
ISBN 1 86320 003 7

This technical paper reports a case study in Niger in which measurements of draft force requirements were made for carts with pneumatic or steel wheels and roller or wooden bearings. A systematic approach is made to describe the factors influencing the draft animal transport system; general considerations for deign of carts are discussed. Results of experimental work are presented that show the influence of construction parameters (pneumatic tyres vs. steel wheels, roller vs. wooden bearings) on draft force requirements, and the influence of forces under different conditions (laterite, sandy road, field and loose sand). Rolling resistance for each factor combination was also determined. For solid tyres the positive effect of springs for the reduction of peaks in draft force was analysed. The study found that the rolling resistance of wooden bearings was 14 per cent higher than that of roller bearings. Steel wheels required about twice as much draft force as pneumatic tyres under the same conditions. Steel wheels also had a very variable resistance with sharp peaks of draft force requirement. In practical terms this led to vibrations of the cart that

Keywords:
animal power,
carts,
bearings,
wheels,
IMT,
technology

were uncomfortable for both animals and driver. Pneumatic tyres damped the variations and had a more constant force requirement. Springs were found to significantly reduce the variations for steel wheels but were expensive The study found that the rolling resistance on a sandy road was three times that on a laterite roads. Mean resistance off road was 6.3 times that of a laterite road. These large differences suggests that cart designs that are suitable for one situation may not be appropriate for another. This paper is illustrated with diagrams and graphs, and should be of interest to those involved in the selection, design or testing of cart wheels and bearings.

DENNIS, RON (ED) (1996)
Guidelines for design, production and testing of animal-drawn carts. A resource book of the Animal Traction Network for Eastern and Southern Africa (ATNESA)
IT Publications, London; 187p.
ISBN 1 85339 338 X

Keywords:
IMT,
animal power,
carts,
bearings,
wheels,
SSAfrica

This resource book of the Animal Traction Network for Eastern and Southern Africa (ATNESA) derives from a workshop held in 1993 in Harare, Zimbabwe. The first section of the book provides detailed technical information on the design and construction of animal-drawn carts and their components. Later sections discuss harnessing systems, manufacturing, marketing, maintenance, credit schemes and testing procedures for animal-drawn carts. The opinions and recommendations of the workshop working groups are included.

Their suggestions for development priorities include:

- development of an affordable puncture-resistant tyre intermediate between a pneumatic and solid rubber tyre
- improvement of the design of a steel wheel for puncture-resistant tyres
- standardisation of size and type of rolling element bearings and organization of a supply and distribution network
- development of a comfortable harness system for two donkeys
- provision by large-scale manufacturers of maintenance packages
- group bulk purchases by small manufacturers of material, components and spares
- publications of directories of stockists and manufacturers.

The book is illustrated with many photographs, technical drawings and tables of technical specifications.

DESHPANDE, S.D. AND T.P. OJHA (1984)
Monograph on traditional and improved bullock carts of India
Technical Monograph CIAE/84/44. Central Institute of Agricultural Engineering, Bhopal, India; 157p.
Available from: ILO/ASIST Nairobi

This book presents the technical specifications of more than thirty designs of ox carts used in India with photographs or technical drawings of each type. Both traditional and so-called "improved" cart designs are included. The information presented includes usual function, manufacturer, dimensions, load-carrying capacity and draft force requirements. Recommendations are made for modifications to the designs. A review section discusses the geographical distribution of cart use in India and local modifications of standard designs. Appendices give addresses of cart manufacturers and organizations involved in research and development of animal-drawn carts in India. The main value of the book is to illustrate the diversity of bullock cart heritage in India.

Keywords:
**IMT,
animal power,
carts,
India**

DOGGER, J.W. (1990)
*Final report ox-cart testing activities
August 1987 - July 1990*
Animal Draught Power Research and Development Project, Magoye Regional Research Station, Magoye, Zambia; 40p.
Available from: ILO/ASIST Nairobi

Keywords:
**IMT,
animal power,
carts,
bearings,
wheels,
technology,
Zambia**

This report presents the results of on-station and on-farm tests of 19 designs of ox cart used in Zambia. Technical specifications and the test results are given for each design. Tests carried out include measurement of draft force, strength/impact tests and on-farm acceptability and durability. The report discusses the suitability of different wheel, bearing and drawbar types in the light of the on-farm tests. Only the 'Lenco' carts with pneumatic tyres and roller bearings passed both on-station and on-farm tests. The report recommends improvements to each design and suggests that some should be scrapped altogether. This publication is likely to be particularly useful to programmes involved in testing animal-drawn carts, and anyone interested in cart design, manufacture and promotion.

DTU (1994)
Simple low-cost wooden ox cart
Technical Release 24. Animal Cart Programme, Development Technology Unit (DTU), University of Warwick, U.K.; 11p.
Available from: DTU, University of Warwick

Keywords:
**animal power,
carts,
technology,
IMT,
manufacture**

A step-by-step guide to the construction of the body of a wooden ox cart suitable for production by village artisans. Includes pictures, technical drawings and details of necessary tools and materials. Other papers in the series include:

No. 21   Pipe and roller donkey cart axles
No. 22   Pipe and roller axle for ox carts
No. 23   Steel frame and wood donkey cart
No. 25   Light steel and wood donkey cart
No. 26   Low-cost steel and wood ox cart

Some of the designs are prototypes that have subsequently been modified. It is not recommended that these designs be copied blindly, but used as ideas (together with other resources) by those working to produce simple, low-cost technologies.

FAO (1983)
Carts for draught animals
FAO, Rome, Italy; 28p.
Available from: FAO/AGAP

This booklet and accompanying filmstrip contains photographs of various models of animal-drawn cart from Burkina Faso. These include a box cart, flatbed carts and a three-wheeled cart. All carts have steel frames, pneumatic tyres and roller bearings. The photographs have brief, non-technical annotations and depict many aspects of cart use and maintenance including correct load placement and maintenance of bearings. It would be a useful resource book for extension workers and those training them.

The booklet is available in English and French from the Distribution and Sales section of the FAO, Rome.

FAO (1994)
Draught animal power manual:
A training manual for use by extension agents
FAO, Rome, Italy; 240p.
Available from: FAO/AGAP

This training manual is designed to be used in the preparation of training and extension courses. It contains practical, non-technical information in words and pictures on the use, training and care of oxen and donkeys for a variety of operations. It is divided into short modules, each dealing with a particular aspect of training and use of draft animals and associated equipment. Chapter 2 describes ways of harnessing and training oxen and donkeys using yokes, collars and breast-bands with simple guidelines for making harnesses. Chapter 5 describes different types of cart and gives guidelines for their maintenance. A section on the use of donkeys as pack animals discusses the merits of pack saddles, baskets and simple packs with recommendations for comfortable loading. The manual also discusses general care of animals and includes notes for training course instructors. Spanish and French versions may be produced (enquiries to FAO-AGAP).

Keywords:
**transport,
animal power,
carts,
donkeys,
harnessing,
pack transport**

FIELDING, D. (1988)
'Pack transport with donkeys'
Appropriate Technology Vol. 15 No. 3: 11-13.
ISSN 0305-0920

Keywords:
IMT,
animal power,
donkeys,
pack transport

This short, non-technical paper discusses the advantages and disadvantages of using donkeys for pack transport, and provides a useful introduction to the topic. It notes that donkeys and pack saddles are cheap, donkeys can negotiate steep and irregular terrain and can be effective where access is limited. Thus despite their lower load-carrying capacity compared with other animals there are many situations in which their use is appropriate and economic. The paper reviews criteria for selection of donkeys for pack transport, citing large size, short and straight backs and good quality hooves as particularly desirable features. The design of pack saddles and possible problems with their use is discussed with emphasis on the need for correct loading to avoid injury to the animal and increase efficiency. The paper concludes that pack transport by donkeys is an appropriate technology for the tropics that should receive more attention by transport workers.

GATE (1989)
Transport for the poor: GATE questions, answers, information 1/89 special issue
Eschborn, Germany; 59p.
ISSN 023-2225
Available from: GATE

An issue of the GATE series focusing on transport for the poor, with articles on general rural transport issues, gender and transport and the role of bicycles in rural transport. The editorial focuses on transport for the poor: basic need or means to an end. Articles relevant to animal power include:

Keywords:
transport,
animal power,
carts,
gender

HARISSON, PETER AND JOHN HOWE
'Measuring the transport demands of the rural poor: Experience from Africa.'
pp: 3-6.

A study of the transport needs of poorer sectors of the populations in Ghana and Tanzania produced the following findings: the transport activities of a rural household in Tanzania occupy 2600 hours per annum and involve a load-carrying effort of 100 tonne-kilometres. The figures for Ghana are 4800 hours per annum and 200 tonne-kilometres. Most transport is effected by women, on foot. And most trips are undertaken to meet agricultural requirements and essential domestic needs. The study was based on selected households.

IMMERS, BEN H., ERNST J. MALIPAARD AND MICHEL J.H. OLDENHOF
'Rural transport in developing countries: A case study of Western Province, Zambia.'
pp: 7-11.

Keywords:
transport,
animal power,
carts,

This article deals with the transport problems of people living in rural areas of developing countries. As the majority of these people are small-scale farmers, attention is focused in particular on the relationship between agricultural activities and transport. To alleviate existing transport problems, attention should be paid first of all to the low variety and availability of intermediate means of transport. The research findings also indicate that agricultural transport is strongly influenced by the way farming households are organised and by the extent of government involvement.

STARKEY, PAUL
'Animal-drawn transport in Africa'
pp: 13-18.

Keywords:
IMT,
animal power,
carts,
poverty,
Zambia,
Tanzania,
Ghana

Two-wheeled carts are the major form of animal-drawn transport in Africa. Sledges without wheels are simple but cause erosion, while four-wheeled carts are complex and expensive. Neither solid wooden wheels nor wooden-spoked wheels have spread in sub-Saharan Africa. Carts with steel wheels have been adopted in some countries, but bearings have often caused severe problems. Artisans can use old vehicle axles to make useful carts. The combination of roller bearings and pneumatic tyres has been particularly successful in West Africa where the social and economic benefits appear to justify the capital cost and the inevitable puncture repairs.

HELSLOOT, HANS, HENRY SICHEMBE AND KENNETH CHELEMU (1993)
Animal powered rural transport in Zambia: Prospects and constraints for development.
Instituut voor Mechanisatie, Arbeid en Gebouwen (IMAG-DLO), Wageningen, The Netherlands; 66p.
Available from: ILO/ASIST Nairobi

Keywords:
IMT,
animal power,
carts,
Zambia,
economic analysis

This case study reviews the current status of rural transport in Zambia in the context of the prevailing political and socio-economic environment, noting that most rural transport currently involves headloading by women. The report analyses the economics of the use of ox carts by smallholder farmers and the current manufacturing and repair facilities in Zambia. It concludes that there is potentially a high demand for good quality ox carts based on roller bearings and pneumatic tyres and a substantial demand for repair facilities. Recommendations for future policies are made including collaboration between large, medium and small-scale manufacturers and establishment of a network for marketing and distribution.

JONES, PETA A. (1991)
Training course manual on the use of donkeys in agriculture in Zimbabwe
Agritex Institute of Agricultural Engineering, Borrowdale, Harare, Zimbabwe; 81p.
Available from: ILO/ASIST Nairobi

Keywords:
IMT,
animal power,
carts,
donkeys,
harnessing,
pack transport

This manual contains detailed practical information on the selection, management, and training of donkeys as well as an overview of their uses and different types of equipment. A section on harnessing for transport describes the theory and practice of the use of swingles and eveners with breastband and collar harnesses for pulling carts. The use of a breechstrap so a donkey can restrain a loaded cart going downhill is described. Different packing systems including bags, pack saddles and panniers are described with recommendations for their construction and use. The design features of equipment that are necessary to prevent injury to the donkeys is emphasized. The manual is designed for extension workers and those training them and includes a plan for a training course. The manual is illustrated with line drawings. Although the manual is an initial draft that would benefit from editing and updating, it is still a valuable resource.

MÜLLER, HERBERT
Oxpower in Zambian agriculture and rural transport: Performance, potential and promotion
Socioeconomic Studies in Rural Development Vol. 65, Edition Herodot, Rader Verlag, Aachen, Germany; 151p.
ISBN 3 922868 40 1; ISSN 0175-2464

This interesting case study on rural transport in Zambia, is a published version of an MSc thesis for the University of Warwick.

In the thesis, draft animal power is examined in the context of Zambian agriculture. Particular emphasis is placed on rural transport and improvements to ox cart transport in Zambia. Interest by government and development agencies in the promotion of animal power for agriculture has grown sharply in Zambia in the last five years. Hopes focus now on the 'emergent farmer' to decrease the country's dependence on food imports. To strengthen the emergent farmers group a large number of subsistence farmers are required to 'emerge'. Ox power is necessary to make the transition from subsistence to emergent farmer. Transport problems have been identified as a major constraint on increasing the number of emergent farmers.

A questionnaire survey was carried out among emergent farmers and ox cart owners in Zambia to establish their actual transport requirements. It was found that ox carts are the appropriate solution to these farmers' transport problems since they are economical for small farmers, well-suited for the type of roads used and adequate in their speed, capacity and range of operation. It is shown how ox carts can increase the area suitable for emergent farmers from the immediate neighbourhood of markets and depots to a 20km radius. A cart specification has been worked out and a design example is included.

In an appendix it is shown how outside agencies have influenced speed and degree of adoption of ox power. Proposals are made for the propagation of animal power and improvements in the manufacture of ox carts to further speed up the process of oxenisation.

Keywords:
IMT,
animal power,
carts,
household
survey

STARKEY, PAUL (1989)
Harnessing and implements for animal traction
GATE/GTZ, Eschborn, Germany; 244p.
ISBN 3 528 2053 9

Keywords:
IMT,
animal power,
carts,
donkeys,
harnessing,
wheels,
bearings,
technology

This book reviews equipment and harnessing systems for animals that are in use around the world. Chapter 8 discusses equipment for transport using draft animals including pack equipment, sledges and carts with comments on their advantages and disadvantages. Wheel and bearing types for carts are discussed and compared. The author finds that wooden spoked wheels are complex to manufacture whilst solid wooden wheels are heavy and unfashionable. Steel-spoked wheels are easier to manufacture and maintain but are stiff and transmit shock through the cart. Carts with pneumatic tyres are most popular despite the problems of punctures. A review of bearing types concludes that despite their higher initial cost, roller bearings are the most reliable and most popular among farmers.

A section on further sources of information gives references to relevant publications and addresses of organizations with a particular interest in animal-powered transport equipment..

Other chapters include discussions of harnessing systems relevant to animal-powered transport. The book is extensively illustrated with photographs and drawings.

Copies of the book in English or French are normally available free-of-charge to organizations in developing countries, and can be requested from GATE/GTZ

STARKEY, PAUL, E. MWENYA AND J. STARES (ED) (1994)
Improving animal traction technology. Proceedings of the first workshop of ATNESA, held on January 18-23, 1992, Lusaka, Zambia.
CTA, Wageningen, The Netherlands; 490p.
ISBN 92 9081 127 7

Keywords:
technology,
SSAfrica

The book derives from a workshop of the Animal Traction Network for Eastern and Southern Africa (ATNESA). Several papers relate to animal-powered transport. The subjects covered include a general analysis of animal-related transport issues, case histories and technical papers. The titles and abstracts of the most relevant papers are given below. The first three papers mentioned are likely to have the greatest interest to most readers: the first two provide useful overviews, while the third provides an interesting case history from Zambia.

ANDERSON, MARY AND RON DENNIS
Improving animal-based transport: Options, approaches, issues and impact
pp. 378-395.

The use of draft animals for rural transport is an important complement to their use in agriculture. The movement of agricultural and subsistence goods is a major burden in time and effort for rural households. The majority of movements take place at farm and village level, often by walking. The use of animals can improve the efficiency of transport, alleviating constraints on farm productivity and aiding agricultural development. However, the potential role of animal-based transport is still largely unrealized in eastern and southern Africa.

Keywords:
**animal power,
IMT,
carts,
SSAfrica**

This paper discusses the options for animal-based transport. Carts have the greatest potential for improving rural transport, although smaller farmers may not be able to afford them. There is a large unsatisfied demand for carts in the region resulting from problems in production, primarily the limited availability of materials and components, particularly good quality wheel-axle assemblies. The issues involved in improving the production of carts are considered and an integrated approach is recommended to improve the supply of materials and critical components to rural workshops which would construct and assemble carts. It is anticipated that this would develop an effective infrastructure for supply and maintenance of carts and provide carts to farmers at minimum cost.

Affordability and profitability of animal-based transport are key issues in its wider dissemination. Experience from many parts of Africa suggests that the availability of credit facilities is of great importance to successful dissemination programmes. The issue of access to transport facilities by women is of major significance in improving the impact of these programmes.

DENNIS, RON AND MARY ANDERSON
Improving animal-based transport: Technical aspects of cart design
pp. 396-404.

Keywords:
animal power,
IMT,
carts,
donkeys,
technology,
SSAfrica

Transport of goods in rural areas in Africa is carried out mainly on foot, imposing a major burden on rural households in terms of both time and effort. Animal-drawn carts can considerably reduce this burden and so improve the productivity of small farmers, but their use is still quite limited. Poor availability and/or high cost of materials and components both constrain production and cause wide variations in the design and quality of construction of carts. One means of improving the quality of carts is to increase awareness of good design practice and of design features that have proved successful in practice. This paper reviews the various issues of cart design and compares some of the available options. It attempts to establish a sound technical base for cart design. The paper concentrates on design of two-wheel carts for oxen and donkeys.

LÖFFLER, CHRISTIAN
Transfer of animal traction technology to farmers in the North Western Province of Zambia
pp. 354-359.

Keywords:
animal power,
IMT,
carts,
Zambia

This paper presents the case study of the introduction of animal draft power in the North Western Province of Zambia within the framework of an Integrated Rural Development Programme (IRDP) sponsored by the German Agency for Technical Co-operation (GTZ). Following a brief analysis of the natural and socio-economic conditions in North Western Province, the paper highlights the specific 'oxenization' approach of IRDP, which combined individual ownership and joint use of work oxen. Special attention is paid to the methods applied during the implementation of the work oxen component in order to make this technological innovation socially and economically viable. Finally, empirical data from various monitoring surveys are presented and analysed to give an assessment of the impact of the IRDP Work Oxen Project in North Western Province.

Other selected chapters of the reader are:

JONES, PETA A.
'A note on a donkey harnessing problem and innovation in Zimbabwe'
pp. 426-427

MUJEMULA, FELIX K.
'Improving animal-drawn transport technology in Tanzania: work on ox carts and bearings'
pp. 414-417

ORAM C.E.
'Low technology rolling element bearings for animal-powered transport and equipment'
pp. 428-434

OUDMAN, LUURT.
'Work on animal power harness technology in Kenya'
pp. 422-425

VROOM, H.
'Rural transport in Zambia: The design of an ox cart which can be produced in rural areas'
pp. 418-421

WIRTH, JOSEF
'Design, adaptation and manufacture of animal-drawn carts'
pp. 405-413

4. RURAL TRANSPORT SERVICES

BEENHAKKER, HENRI WITH S. CARAPETIS, L. CROWTHER AND S. HERTEL (1987)
Rural transport services: A guide to their planning and implementation
IT Publications, London; 379p.
ISBN 0 946688 58 3

Keywords:
planning, infrastructure, transport aids, accessibility, cost evaluations, transport services

This book considers the problem of providing maximum access to transport services and to roads for the rural populations of developing countries when funds are limited. Access, by promoting social intercourse and opening up markets, is crucial for economic and social development. It connotes the ability to travel and to transport goods and thus includes both transport (or rural access) infrastructure as well as transport modes or aids. Rural access infrastructure includes feeder roads (that connect rural areas to main road networks) and tracks, trails and paths (that provide local access to those feeder roads). Conventional transport aids –trucks, buses, etc.– are often of less importance to rural communities than traditional transport aids, including head baskets, carts and pack/draught animals.

The authors assert that it is only possible to solve transport and access problems when both the infrastructure and the aids are available and are considered together in order to meet specific transport requirements. The text attempts to describe comprehensively a multifaceted approach to the provision of rural transport infrastructure and services and to suggest a variety of solutions to specific problems.

The book addresses the needs of decision makers, planners, engineers and technicians in developing countries, as well as aid agencies, NGOs and consultants. It offers a rationale for modifications to planning and implementing rural access infrastructure, and guidelines and procedures for identifying transport problems and their solutions. It is intended to be used innovatively in a variety of country situations; although it is not country-specific, data and examples are included to illustrate specific points.

The book is organised in a way to cover all the major topics. These include accessibility; construction; cost evaluations; drainage; economic analysis; infrastructure improvements; institutional issues; local labour; maintenance; objectives, policies and strategies; planning; spot improvements; and transport aids.

BIGGS, S., A. KELLY AND G. BALASURIYA (1993)
Rural entrepreneurs, two wheel tractors and markets for services: A case study from Sri Lanka
Discussion paper No. 242, The School of Development Studies, University of East Anglia, Norwich; 36p.
ISBN 1 898285 85 3

This discussion paper describes the results of an exploratory study to analyse rural mechanization processes and policies. The paper reviews agricultural mechanization policy in Sri Lanka and the research specifically looks at the rapid spread of two-wheel tractors since 1977. It examines the reasons for their diffusion and the distribution of benefits during this period of technical and institutional change. The research analyses the role played in this process by economic policy and by different institutional actors in the private, government and non-government sectors.

The paper is divided into six parts and includes an introduction to Sri Lankan agriculture and labour markets, and rationale for the introduction of tractors after the second World War. The third part of the paper concentrates on economic reform after 1977 and the impact that this had on the spread of two-wheel tractors. The fourth part uses the two wheel tractor as a case study to examine in detail the reasons for its rapid adoption. Part five of the paper looks at the policy implications and part six discusses the role of a 'dynamic interactive model' in helping to understand past rural mechanization processes.

Keywords:
**Sri Lanka,
tractors,
rural
entrepreneurs,
mechanization,
policy,
planning,
transport
services**

The rapid spread of two-wheel tractors in Sri Lanka is attributed to economic liberalization and the innovative capabilities of small-scale rural entrepreneurs who had experience and access to a wide variety of skills. These entrepreneurs included vehicle operators, repair and maintenance artisans as well as large-scale urban machinery manufacturers and importers. It was the skill and forward planning of these entrepreneurs which meant that early fears about insufficient backup services and the poor availability of spares for the imported tractors were not realized. It is concluded that policy studies are required to assess past and present rural mechanization policies and their impact on poverty reduction.

BYRNE, H.M. AND D.J. SAVAGE (1984)
Thuchi-Nkubu road evaluation study – Rural passenger travel demand: Preliminary analysis and results
Overseas Centre Working Paper No. 156, TRL, Crowthorne; 61p.
Available from: TRL

Keywords:
Kenya, elasticities, accessibility, transport services, household survey

This working paper reports the results of a survey of personal trip making in the Southern part of the Meru District, Kenya. The aim of the survey was to collect data with which to assess the impact of road improvements on the travel patterns of the people living around the road. Personal travel information was collected from households in representative village areas. The data was collected using travel logs for each household member over a months period and a travel needs questionnaire.

Background socio-economic information is presented for the sample households and includes demography, land holdings, sources of income, availability of credit and the ownership of transport. The subsequent analysis provides information on the pattern of monthly journey making, the determination of the price elasticity of demand for journey making and for travel, and the identification of project benefit recipients. Regression relationships are derived for factors thought to affect demand for journey making including household size, frequency of transport service, income, cost of transport and trip distance.

An analysis is also presented on the impact of household income and accessibility on trips made for specific purposes. The analysis includes trips for work, school, trading, personal administration, social, health, shopping and banking. The results show the impact on these various types of trip making according to distance from district centres and cost of travel.

The working paper concludes that the road improvement is likely to increase opportunities for the individual to travel by matatu (local paratransit vehicles). However, price elasticities for travel appeared to be low at -0.4 (i.e. a 1 per cent reduction in transport costs will lead to a 0.4 per cent increase in demand for travel).

An updated version of this report will soon be available from TRL.

CARAPETIS, S., H. BEENHAKKER AND J. HOWE (1984)
The supply and quality of rural transport services in developing countries: A comparative review
World Bank Staff Working Paper No. 654, The World Bank, Washington, D.C.; 52p.
ISBN 0 8213 0390 2

This World Bank Working Paper contains the results of a study carried out through a comparative review of the supply and quality of rural transport services. The main means of data collection was a series of short field surveys in the Philippines, Indonesia, Sri Lanka, India, Sierra Leone and Tunisia during 1981 and part of 1982. The review also includes relevant data and information from other sources.

Keywords:
**transport services,
rural roads,
policy,
planning**

The general issue examined in this report is the extent to which transport services that should complement investments in rural roads can be presumed to materialize and satisfy movement demands efficiently. A related concern is whether those services that do materialize are physically and financially available to all rural people.

The report is structured around five themes. These include rural transport infrastructure, the vehicles as well as the various travel and transport aids available and in use, and discusses the general direction and focus of past and current investment programmes, the implications for the future and prospects and options for change. There is a discussion of the transport disenfranchised, their plight and the prospects and options for reaching them. Government policies concerning rural transport matters and government interventions are considered. The need for a new perspective and approach to planning and formulating rural transport investment projects and programmes is emphasized. Finally there are ideas and suggestions for government attention and action and suggestions for the role of the World Bank in the matter.

The study reveals a need for new perspectives in planning rural transport investments which require a better understanding of the real transport needs and problems of small farmers. These needs are often for personal travel and for moving small loads over relatively short distances. The present emphasis in rural road investment, of improving existing roads and extending them by relatively marginal distances, is frequently ineffective in the economic and social development of low-income farmers.

DELAQUIS, MICHÈLE (1993)
Vehicle efficiency and agricultural transport in Ghana
MSc thesis, University of Manitoba, Canada; 193p.
Available from: ILO/ASIST Nairobi

Keywords:
Ghana,
trucks,
costs,
transport services,
transport efficiency

This MSc thesis studies the vehicle operating costs associated with the transportation of agricultural commodities in Ghana using Kumasi and the Ashanti Region as a case study. The present state of the agricultural sector is described in terms of three interactive systems: the transport system, the agriculture system, and the flow pattern of vehicles and commodities. A survey is used as an information base to construct a Truck Operating Cost (TOC) model based on average actual operating conditions. The TOC model is expanded to include costs under three theoretical operating conditions: enforced loading conditions, maximum vehicle utilization, and increased fuel efficiency.

Three options identified as potentially beneficial to the transport industry and the Ghanian economy, are presented and evaluated: using larger vehicles, maximizing vehicle utilization, and increasing fuel consumption. Evaluation takes into consideration the effects of implementation on the groups involved; these being the producers, transport owners and operators, transport organizations, and the government.

Recommendations put forth suggest that the Ghanian government in co-operation with the Ghanian Private Road Transport Union (GPRTU) and lorry park operations institute the following programs and policies: a) to enforce registered loading allowance; b) encourage higher vehicle utilization by controlling the number of vehicles registered and ensuring adequate service on all routes; c) encourage the use of larger vehicles. Union porters consolidate and load all vehicles at GPRTU lorry parks.

It is further recommended that the Ghanian government examine the benefits of using foreign aid to effect such fleet and operational changes rather than focus on capital intensive infrastructure improvements in order to improve transport efficiency. Low vehicle utilization and high overloading are major determinants of the present transport cost structure in Ghana. Low utilization is caused by long times between loads and long periods out of service due to repairs.

ELLIS, SIMON (1996)
The economics of the provision of rural transport services in developing countries
PhD thesis, Cranfield University; 325p.
Available from: TRL

This PhD study has attempted to address the issue of unreliable and often expensive rural transport which hinders agricultural development in the rural areas of developing countries. The aim of the project has been to examine the supply of vehicle services, with particular regard to the consequences of low vehicle numbers and diversity in rural Sub-Saharan Africa, and the effect that this may have on transport charges, competition, operator efficiency and service reliability.

Keywords: transport services, competition, costs, low-cost vehicles, planning, IMT, SSAfrica

Surveys were undertaken in Thailand, Sri Lanka, Ghana, Zimbabwe and Pakistan. Data were collected on vehicle operating costs (VOC) and performance for a wide range of commonly used rural vehicles. These included human porterage and non-motorized vehicles such as bicycles and animal transport, as well as motorized vehicles such as conventional trucks and pickups, agricultural tractors and simple engine-powered vehicles. Analysis of the data demonstrated large differences in the VOC and transport charges for rural transport services between the generally efficient systems in the Asian countries and the inefficient ones in the African countries studied. This finding formed the foundation for the development of a framework which identified the relationships between transport charges, VOC, and factors relating to the operating environment.

The framework was used to develop a knowledge based computer programme or expert system which has been named the "Rural Transport Planner" (RTP). The RTP is designed to aid in the selection of appropriate transport vehicles and to make recommendations on interventions that may improve the efficiency of rural transport systems. The programme contains a database of 47 rural vehicles, a model for the calculation of VOC and total distribution costs, and a knowledge base to improve the selection of rural transport interventions.

It is concluded that there are vehicle shortages in much of rural Sub-Saharan Africa that in part can be explained by government policy, lack of competition, poor marketing opportunities and ineffective agricultural extension services.

Ellis-Jones, Jim and Brian G. Sims (1995)
A survey of rural transport in the Guinope Municipality, Honduras
Unpublished report, Silsoe Research Institute; 34p.
Available from: ILO/ASIST Nairobi

This report is a component of a wider Overseas Development Administration (ODA)-funded project examining rural transport vehicles. This study is based on a rapid rural appraisal of Guinope municipality in Honduras. The data was collected with the use of semi-structured household and vehicle operator questionnaires, and supplemented with information from key informants.

Keywords:
Honduras, transport services, policy, IMT, household survey, costs, maintenance

The Guinope municipality is a mountainous area typical of rural conditions in the country where 80 per cent of the population are engaged in agriculture. The survey examined the availability of transport vehicles, their operating characteristics, operating environments and the factors affecting transport costs. The study found a wide variety of animal and motorized modes of transport operating according to the infrastructure and demand available.

The report covers the effect that various factors have on the efficiency and costs of transport operations in the municipality. These factors include government policy, the role of transport owners associations, repair and maintenance problems, road conditions, the impact of traders and access to alternative income sources. Vehicle operating costs are provided for a wide range of vehicles from wheelbarrows, pack animals, animal carts and conventional vehicles. These are supported by the vehicles respective transport charges.

The study concluded that although the availability of transport services was good there were a number of issues that required further consideration. There was a shortage of skilled mechanics because of the proximity of better labour markets, particularly in the USA, and therefore local skills need upgrading and retaining within Honduras. The high costs and strict conditions applied to formal credit make these sources of credit unavailable to the majority of rural transporters which leads to a reliance on informal sources. The rural infrastructure is in a very poor condition and cost-effective maintenance is urgently needed.

NATIONAL TRANSPORT RESEARCH CENTRE (1989)
Survey of agricultural tractors
National Transport Research Centre, Government of Pakistan, Pakistan; 40p.

QURESHI, M. AND S. KHAN (1990)
Role of tractor trollies in rural transportation
NTRC, Government of Pakistan, Pakistan.
Available from: ILO/ASIST Nairobi

Keywords:
Pakistan,
tractors,
trailers,
policy,
safety,
transport
services

There has been a phenomenal growth in the population of tractors in Pakistan during the 1980s. Although the tractor is basically an agricultural implement, it is being invariably fitted with a trolley (trailer), and widely used for the purpose of transporting agricultural as well as non-agricultural produce. The tractor trolley is rapidly replacing animal drawn vehicles and trucks in rural areas. However, the tractor trollies have axle loads which are much heavier than trucks and are being involved in a growing number of accidents. The purpose of this report is to determine the role of tractor trollies in rural transportation and their social and economic impact, both as a means of transportation and on road safety.

The report covers the impact of government policy on the ownership of tractor trollies. Favourable import duties and taxes along with concessional credit have made tractor operations very profitable in comparison to other modes of transport and as such they have been used extensively as road haulage vehicles. The impact that these tractors have had on agricultural mechanization has been far more modest. However, the manufacture of trollies has created much needed rural employment and there is rapidly increasing demand.

It is concluded that although the tractor trolley plays a very important role in rural life in Pakistan there needs to be greater regulation over their use and particularly over the design of the trollies. Many of the existing designs are unsafe for the loads being carried and recommendations are made for the necessary improvements. It is also recommended that driver training programmes are set up to educate operators on safe driving techniques.

This work follows an initial report of the results of a survey of tractors and tractor trollies; including vehicle numbers, ownership, utilization, costs and productivity changes.

PLUMBE, A.J. AND H. BYRNE (1981)
The role of the agricultural tractor in road haulage in Sri Lanka
TRRL Laboratory Report 1007, TRL, Crowthorne; 16p.
ISSN 0305-1293
Available from: TRL

This report presents the results of a survey of freight movements in Sri Lanka and clearly indicates the importance that both four- and two-wheeled tractors make to road haulage. While some machines have been acquired by operators to use exclusively for this purpose, others are employed in this role only when not needed for soil cultivation. The report describes the geographical distribution of ownership, the scale of employment of tractors in this role and shows that before November 1977 road haulage was a profitable business for the owners of machines. After this date when the import duty was substantially increased the report explains that the position was changed with owners of new machines being unlikely to cover their capital and running costs from revenue.

The costs of two- and four-wheel tractor operations are set out and these are set against revenues in order to calculate the expected break-even positions for tractor operations. The results indicated that the profitability of tractor operations were very sensitive to increased import duty. Before the 1977 budget the break-even position for four wheeled tractors was about 90 days work per year but this increased to 150 days after the budget. There was a similar situation with two wheeled tractors whose break-even positions moved from 35 days work per year to 100 days per year.

It is concluded that tractors and trailers play an important role in the transportation of loads weighing around three tonnes over distances of about 10 kilometres. Rural entrepreneurs have found tractor operations very profitable and have hence used tractors predominantly for road

Keywords:
Sri Lanka,
tractors,
transport services,
costs,
policy,
low-cost vehicles

haulage operations. Following the 1977 budget it is expected that transport charges will rise and possible difficulties with replacing tractors at current market prices.

PLUMBE, A.J. AND D.J. SAVAGE (1981)
Bullock cart haulage in Sri Lanka
TRRL Laboratory Report 1006, Transport Research Laboratory, Crowthorne; 18p.
ISSN 0305-1293
Available from: TRL

Keywords:
Sri Lanka,
animal power,
bullock carts,
costs,
transport services;
technology

This report presents the results of surveys conducted in Sri Lanka from postal questionnaires sent to local government authorities, interviews with bullock cart owners, peri-urban roadside surveys and rural cross-roads surveys. The characteristics of bullock cart haulage are described in terms of the goods they carry, payloads, speeds and trip distances, utilization and bullock cart charges.

The components of bullock cart operating costs are detailed as well as their expected revenues and estimated profitability. Some carts are used predominantly for hire while others are used just for the farmers own use. It is calculated that bullock cart owners having only one cart and no other business enterprises fall between the third and fourth lowest decile of income earners in the country.

The report concludes that while the basic design of the bullock cart has changed little over time, and could well be improved, it is easily maintained and well suited to its role in satisfying the demand for goods transport at the most local level and in small units. However, there is concern that the traditional design of the bullock cart is not as efficient as it could be. For example, the narrow steel rimmed wheels cause extensive damage to the road pavement, the yoke is inefficient and often carts are badly balanced with small load capacities.

It appears that tractors and trailers and bullock carts serve the same market but that the tractors have started taking over much of the business. However, it is suggested that bullock carts have the advantage that they are locally manufactured and therefore not reliant on expensive imported spares.

SRISAKDA, LAMDUAN AND SOMPONG CHIVASANT (1992)
Study of the 'Itaen' rural transport services in northern Thailand
World Bank Working Paper 6, Report prepared on behalf of IT Transport Ltd. for World Bank SSATP. IT Transport Ltd., Oxford, and ILO, Geneva; 45p.
Available from: World Bank SSATP

This report is one of a series of village-level transport and travel surveys carried out under the Rural Travel and Transport Project of the World Bank Sub-Saharan Africa Transport Programme. This case study examines the use of "Itaen" vehicles in Northern Thailand. The Itaen has a simple fabricated steel chassis to which are assembled second-hand suspension and drive-train components. It uses a single-cylinder air-cooled diesel engine, usually 8-12hp. The Itaen is produced by small industries and the complete vehicle costs about one-third the price of a conventional Japanese one tonne pick-up.

Keywords:
**Itaen,
Thailand,
manufacture,
costs,
efficiency,
technology,
low-cost
vehicles,
transport
services**

The report covers the history and development of the Itaen in Thailand and the extent of its use in the different regions. The characteristics of the manufacturing industry are described together with the vehicle's technical specifications, purchase costs and performance. A section of the report describes the study areas, characteristics of the vehicle owners and the passenger and agricultural transport services that are provided by the Itaen. The profitability of these operations is also examined together with a section on the simple regulations governing the operation of these vehicles.

The report concludes that there are a number of factors which have contributed to the success of the Itaen, these include: the availability of all-weather roads; relatively flat terrain; high demand for the transport of produce; relatively high incomes and therefore the affordability of the vehicle; an infrequent bus service in the rural areas; an acceptable quality of service; the ability to manufacture, assemble, maintain and repair the vehicle; and the availability of credit. It is considered that the only negative impact of the Itaen is the lack of safety: this is particularly the case when the vehicles are operated on the highways.

5. RURAL TRANSPORT PLANNING

AFFUM, JOSEPH AND FARHAD AHMED (1995)
Use of GIS as a decision making tool in integrated rural transport planning
University of South Australia; 10p.
Available from: ILO/ASIST Nairobi

Keywords:
**IRAP,
planning,
prioritization,
assessment
methods,
Bangladesh**

This paper details the use of Geographical Information Systems (PC ARC INFO) for rural accessibility planning in Bangladesh. The authors use the software to assess the effects of road improvements on the accessibility of facilities at the ward level. They calculate the absolute (observed) accessibility and the ideal accessibility by using airline (Euclidean) distances. The ratio of absolute impedance and ideal accessibility, called relative accessibility, can be used as an indicator for the performance of the existing transport infrastructure. The software produces a ward level map of the relative accessibility, which indicates the potentials of road improvements. In a second step scenarios of road improvements are calculated: the change of the absolute accessibility depicts the generated absolute improvements of accessibility. After the improvement the ratios of relative accessibility show more equal levels throughout the district.

The authors give some practical hints how to program the software. The methodology is only used to assess the effects of road improvements on the accessibility of central facilities. It would be interesting to research if the software can be used as well for the assessment of accessibility improvements on the village level.

BARWELL, IAN AND JONATHAN DAWSON (1993)
Roads are not enough: New perspectives on rural transport planning in developing countries
IT Publications, London; 79p.
ISBN 1 85339 191 3

This book is an introduction to the appropriate approach towards transport planning in rural areas of developing countries. It was published as an introduction to the issues that the International Forum for Rural Transport is promoting. The authors give an overview of the research that has been done in the last decade and discuss its implications on transport policy. They argue that the focus on rural roads is not enough to satisfy rural transport demands. A more holistic approach is needed, which takes into account the access and mobility needs of rural households.

Keywords: **transport planning, community level surveys, appropriate transport**

The first chapter traces the evolution of transport theory and policy from a bias on roads and cars in the 1980s towards a broader view in the 1990s. The second chapter describes the methodology and results of four community-level transport surveys conducted in Tanzania, Ghana and in the Philippines. The main transport activities of rural households in terms of trips, time, and tkm are mainly in and around the village and they are undertaken mainly by walking. Domestic purposes such as water and firewood collection and production related trips dominate the transport patterns, while trips to markets comprise a small share. Women carry the biggest share of the household's total transport burden. The fourth chapter describes the changes of transport interventions in order to satisfy the access demands of rural households. The authors call for the increased use of Intermediate Means of Transport (e.g. carrying devices, human powered vehicles, animal power, low-cost motor vehicles and boats), the development of local infrastructure (paths, tracks and low cost roads), the improvement of transport services and the implementation of transport-avoiding measures by regional planning.

CONNERLEY, ED AND LARRY SCHROEDER (1996)
Rural transport planning
Approach Paper, SSATP Working Paper No. 19,
The World Bank, Washington D.C.; 48p.
Available from: World Bank SSATP

Keywords:
**planning,
theory,
decentralization,
participation,
financing,
SSAfrica**

This Approach Paper was prepared under the Rural Travel and Transport Programme of the World Bank to improve local transport planning procedures in SSAfrica. The authors take a wider approach towards rural transport by combining the concept of rural accessibility with a strong economic bias.

Planning is defined as a process of efficient and transparent allocation of resources as an integral part of national government. The emphasis on efficiency signifies the goal to maximize net benefits from access by keeping costs as low as possible and to ensure that the facilities provided are those demanded. Therefore the concept of 'public service industry' is introduced as a method to analyse the economy of the provision and production of public services. While the government plans the provision of these services, the production can be more efficiently organized by non-government actors, e.g. private sector, communities, NGOs etc.

The authors criticize a number of current ideas and approaches in rural transport planning. In their view, basing plans on defined access needs instead of on transport demands does not reflect what users are actually willing and able to purchase. Another shortcoming is the missing analysis of the affordability of transport interventions in rural communities more often than not resulting in the over design of transport infrastructures. The concept of 'creating a sense of ownership' is a fiction, because real ownership involves enforceable claims over control, possession or use of assets.

The document gives an overview of the assignment of responsibilities for rural transport planning on the national and local level. The authors propose:

- strengthening of decentralized government capacities
- intersectoral co-ordination of local plans
- simplification of planning techniques

- widespread use of a participatory planning process
- cost sharing of beneficiaries.

The book contains nine practical examples of planning and institutional issues in SSAfrica.

EDMONDS, GEOFF, CHRISS DONNGES, AND NORI PALARCA (1994)
Planning for People's Needs
ILO/DILG, Manila.
Available from: ILO/POLDEV Geneva and ILO/ASIST Nairobi

Integrated Rural Accessibility Planning (IRAP) is a local level planning process which is based on the concept that the lack of access of rural people to goods and services is one of the fundamental constraints to their development. IRAP uses access to define priorities for transport investments in relation to water, health, education, energy, agricultural inputs and outputs, roads, markets and transport services. Five papers constitute this series and had been prepared as part of an IRAP project conducted by the ILO in the Philippines.

Keywords: **IRAP, planning, accessibility, Philippines, assessment methods, prioritization, capacity building**

Guidelines on integrated rural accessibility planning
32p.

The guidelines (the first of this series) are intended to provide an introduction to IRAP. It describes the basic philosophy behind this local level planning procedure, how it is applied and finally illustrates what can be expected from it.

After the key features and resource requirements of IRAP are described, the IRAP planning process is outlined. The first step is the data collection, followed by the processing of data and set-up of accessibility profiles and mapping. Accessibility indicators are numeric values, which indicate the ease or difficulty of households to gain access to goods and services. With the help of these indicators a prioritization of interventions is undertaken and action plans developed. The interventions are not only focusing on transport infrastructure, but also on the location of facilities, the provision of transport services and the supply with conveyances.

Because the IRAP process is accompanied by training the outputs are not only a prioritized list of wards, accessibility data bases and maps (1:50 000), but also enhanced local planning capacities. The authors state that the outputs may help as well for provincial and national investment decisions. However, this role might be limited due to the missing assessment of costs in the planning methodology.

The other four papers give practical advice for the planning process:

- *Accessibility data base*
- *Rural accessibility planning and the use of accessibility indicators;* 9p.
- *Rural accessibility mapping;* 12p.
- *Capacity building*

HOWE, JOHN (1990)
Benefits of rural roads: Current issues and concepts
IHE, Delft; 22p.
Available from: ILO/POLDEV Geneva and ILO/ASIST Nairobi

Keywords:
theory,
socio-economic
impacts,
assessment
methods

Howe produced this overview paper for his students at the IHE in Delft. It very briefly discusses various approaches, concepts and theories concerning transport investments in developing countries.

In the first chapter Howe discusses socio-economic effects and impacts of road investments and the role of transport services. On four pages he lists the positive and negative impacts of new roads summarized from USAID evaluations. The next chapter is about the distribution of benefits observed in various projects, the methodologies to assess them and conditions for an improved benefit distribution. Two annexes provide simple assessment methodologies for rural transport. Firstly a trip distribution model is explained which was developed for the estimation of personal travel in rural areas. The second annex discusses indicators for the measurement of mobility and accessibility.

HOWE, JOHN (1983)
Conceptual framework for defining and evaluating improvements to local level rural transport in developing countries
ILO, Geneva; 70p.
Available from: ILO/POLDEV Geneva and ILO/ASIST Nairobi

Keywords: **assessment methods, theory, transport needs, planning**

This report is based on the premise that it is essential to examine possible ways of meeting the basic movement needs of rural communities, given the impossibility in the foreseeable future of extending the conventional road network and transport services to the majority of rural populations. The 'local transport system' can be considered as that which people in rural communities use in their attempts to fulfil basic needs for shelter, food, water, clothing, health services, education and markets for produce. A local system —characteristically of footpaths, tracks and unengineered earth roads on the one hand, and a variety of means of movement from walking through animal-powered and pedal-powered vehicles to simple machines —does exist in most rural areas. Until recently, this end of the transport spectrum has been largely ignored in development programmes.

The report has four main purposes:

- define the specific local level transport needs that should be addressed;
- develop a system of analysis which will lead to the definition of specific project objectives;
- define a system of analysis for identifying the type and scope of interventions in the transport system that will be necessary to achieve those objectives;
- assess the various methodologies that exist for evaluating the effect of the interventions so as to identify the most appropriate.

The report establishes how a basic needs approach to the definition of local level requirements might differ from that of previous transport strategies by considering, first, the relationships between transport and economic and social development that can be implied from past investment patterns and, second, how the planning of local level transport improvements could differ from these relationships if a basic needs approach were adopted. The author claims that the concept of 'basic needs' is an improvement over past methods in that it changes the emphasis from agricultural inputs and outputs to the movement needs of people themselves.

The paper suggests methods for analysing local level transport requirements. It also discusses how the type of transport improvements to meet local level requirements will be identified, and the problem of evaluating interventions in the local level transport system.

Results of studies of local level travel in Asia and Africa, carried out in 1978-1982 emphasize that subsistence-related tasks dominate household travel, that most trips take place around the village/community, that longer trips are motivated by social/welfare considerations, that the total weight/consignment size of goods is small, that a significant proportion of households possess no form of vehicular transport and that there is an overall dominance of simple means of transport (walking, cycling, animal-powered). However, the studies do not indicate how to measure local level travel requirements for practical planning purposes. An annex of the paper provides detailed information on the methodologies, definitions, advantages and disadvantages of the different approaches, objectives and focus of the studies.

STOWERS, J. AND A. TALVITIE (1995)
Highway functional classification study guidelines
The World Bank, Washington D.C.; 80p.
Available from: World Bank Bookstore

Keywords:
roads,
institutions,
policy,
planning,
decentralization,
participation

This World Bank study had the target to define a functional classification of roads to facilitate international comparison of transport networks. The report provides flexible guidelines in order to be useful under the widest range of conditions. Although it focuses mainly on industrialised countries, the methodology may be as well applied in the South.

The concept of functional classification is simple: classify roads into three basic classes, arterial, collector and local. A functional classification takes a larger approach by incorporating the views of people and organizations regarding the function which the network has for their daily lives, the economy and the environment. The classifications can be used for a wide range of purposes in transport and development planning, administration, route guidance, community participation and policy development. The main goal is to find an optimal allocation of government resources.

The first chapter describes theoretical concepts of functional classifications and gives detailed descriptions of the main characteristics of each class. The next two chapters handle the process of conducting a functional classification. First, two examples of a bottom up (USA) and a top down approach (Finland) are presented, before the procedures of classification in rural and urban areas are discussed. In rural areas (1) the population centres are ranked, (2) the arterial system defined and then (3) the collector system identified. The bias to Industrialized Countries explains why this paper does not deal with the classification of local roads, tracks and paths. Possibly a similar methodology can be used to create further sub-classes for these types of rural infrastructures.

THE WORLD BANK (1991)
Republic of Madagascar: Rural road sub-sector strategy
Report No. 9555-MAG, Washington D.C.; 37p.
Available from: World Bank Bookstore

This report presents a practical example about a country strategy for the improvement of rural roads and gives a good overview of the various aspects of rural infrastructure planning. The main goals of the strategy are:

- to reduce the backlog of rehabilitation and maintenance
- to support decentralization
- to enhance the macro-level planning system to improve allocation of resources among regions
- to improve road maintenance funding
- to form a consensus on the appropriate design and construction standards.

After the analysis of the current problems the strategy is developed, which contains the strengthening of institutional capacities, improvement of the planning system, resource mobilization, design and technology for road construction and environmental considerations. The paper does not include infrastructure designed for non-motorized vehicles.

Keywords:
Madagascar, planning, maintenance, financing, decentralization

6. FINANCIAL AND INSTITUTIONAL ISSUES

BALCERAC DE RICHECOUR, ANNE AND IAN G. HEGGIE (1995)
African road funds: what works and why?
SSATP Working Paper No. 14, The World Bank,
Washington D.C.; 28p.
Available from: World Bank SSATP

Keywords:
maintenance,
planning,
roads,
institutions,
decentralization,
financing

This paper presents the lessons learned from road funds installed in ten African countries during the Road Maintenance Initiative (RMI). The RMI approach is described in Heggie (1995). Although most African Road Funds suffer from systemic problems, this review identifies examples of best practice. The review concludes that road funds can work successfully if they have:
- clear objectives
- independent source of revenues mobilized through road tariffs
- effective management of the road fund
- commercial accounting systems.

The paper gives much practical advice concerning all the above mentioned targets.

The main findings are that road funds have to cover their entire expenditure by user charges. An independent board should manage the fund according to sound commercial principles and decide about the level of user charges, which are preferably collected by fuel companies. One of the main lessons learned is that even if maintenance receives the highest priority and no new roads are built, most countries can only afford to maintain a core network. The remainder of the network (mostly rural roads) have to be handed over to lower levels of government or will receive minimum levels of maintenance.

COOK, CYNTHIA, HENRI BEENHAKKER AND RICHARD HARTWIG (1985)
Institutional considerations in rural roads projects
World Bank Staff Working Paper No. 748.,
Washington D.C.; 73p.
ISBN 0 8213 0588 3

This paper is based on the experience of 50 World Bank financed rural road projects. It deals with many practical issues concerning the organization of labour-based road works, but it might be as well consulted for paths and tracks.

The first part considers rural road planning as a part of a wider policy environment, which regards political and economic constraints, the role of the private sector, restrictions to labour-based construction, legal and regulatory requirements, central government planning.

The second part discusses organizational options like implementing issues, inter-agency linkages and structural alternatives. The authors plead for decentralized responsibilities for planning, construction and maintenance of rural roads. In general it is preferable to modify existing institutions rather than creating new ones. Instead of hiring highly skilled staff it is preferable to train existing staff. Hiring local unskilled people on a temporary basis can improve local participation. Task work payments are preferred because labourers can use their time more efficiently. If salaries are too low to attract highly qualified staff, additional incentives (e.g. free housing) are often cheaper than expatriate salaries.

The third part considers local participation during planning, construction and maintenance phase, which increases the sustainability of the project, because local interest is taken into account and local skills, knowledge and labour are used. Best results can be achieved by using a bottom up approach. It is important to involve disadvantaged or underrepresented groups. For maintenance three options can be chosen:

Keywords:
**planning,
institutions,
training,
decentralization,
maintenance,
labour-based,
roads**

self-help labour organized by rural authorities; local contractors, which mobilizes a regional labour force, and the 'lengthman' system, where individuals are responsible for the maintenance of road sections.

HEGGIE, IAN G. (1995)
Management and financing of roads: An agenda for reform
World Bank Technical Paper No. 275, The World Bank, Washington D.C.; 154p.
ISSN 0253-7494

Keywords:
maintenance, planning, roads, institutions, decentralization, financing

Heggie can be regarded as one of the main initiators of the Road Maintenance Initiative (RMI), which was launched by the World Bank in response to the massive road deterioration due to lack of maintenance in SSAfrica. This book describes the main agenda for reform of the African Road sector.

The author proposes the commercializing of African roads by bringing roads under market forces, putting them on a fee for service basis and managing them like any other business enterprise. The salient feature of the reform is the installation of independent road agencies on the various administrative levels. Four basic building blocks should shape the reform:

- Creating ownership by involving road users: installation of a board of directors with representatives of road transport industry, contractors, consultants, farmer's associations and concerned government departments.
- Stabilise the financial flows by mobilizing revenues and linking revenues and expenditures, e.g. by the installation of a road fund.
- Clarify the responsibilities among different government departments and road agencies.
- Improve the management by creating a more businesslike environment: pay the staff adequately, install effective management structures where managers can act commercially, introduce auditing procedures, create contracts between the government and the road agency.

The book supplies many tables, graphs, country case studies and practical examples, and provides in its annexes background information on the economics of road maintenance.

HEGGIE, IAN G., E.O. DONKOR AND CHRISTINA MALMBERG-CALVO (1995)
'Establishing a clear and consistent organizational structure'
Paper presented at the RMI Second Regional Seminar on Management and Finance, Nairobi, Kenya.
Available from: ILO/ASIST Nairobi

The paper deals with the topic of assigning responsibility for roads and road traffic between the various agencies involved in the road sector. It is impossible to improve the management of roads, or strengthen managerial accountability unless each road agency knows what it is responsible for. The paper explores the basic principles applied when assigning responsibilities for roads and road traffic, reviews their application in the African context and suggests ways in which responsibilities might be more clearly assigned between the various agencies.

Keywords:
**institutions,
roads,
SSAfrica,
maintenance,
safety,
management**

The authors provide a good overview of the situation in several African countries. They deal with the problem of management of unproclaimed roads (that is, roads that are not legally assigned to any organization). They examine how this problem has been tackled in Finland and Canada and suggest that the system in Finland could be used as a basis for developing a model for Africa. In Africa, the authors acknowledge that there are a large number of roads that are not proclaimed and the responsibility for the network is divided and ambiguous. For the management of road traffic, the authors recognize that in most African countries the private sector has taken over the responsibility for providing public transport services. In this context, the government must retain the responsibility for regulating the services. The paper recommends different strategies for the allocation of responsibility for more equitable distribution of services, for vehicle safety and for managing traffic.

MALMBERG-CALVO, CHRISTINA (1997)
The institutional and financial framework of rural transport infrastructure
SSATP Working Paper No. 17, The World Bank, Washington D.C.; 86p.
Available from: World Bank SSATP

Keywords:
**institutions,
financing,
maintenance,
paths,
roads,
management**

This approach paper prepared under the Rural Travel and Transport Programme of the World Bank in SSAfrica, focuses on the institutional and financial aspects of rural transport infrastructure, addressing both the lowest levels of the local government network and community roads and paths.

After explaining the context of rural transport, the author describes the major institutional problems related to transport infrastructure: unclear responsibilities, disintegration of the planning system, insufficient maintenance, inadequate local capacities and inappropriate technical standards. Weak local institutions are the underlying causes of these problems.

In the next chapter the author gives a framework for reform which contains questions about ownership and responsibility and the development of a rural transport policy.

The following chapter discusses the institutional framework for managing local government roads. Four organizational options are analysed: (i) central government agency, (ii) contract management, (iii) joint service committee and (iv) private consultants. The author devotes a section of this chapter to the financing of maintenance from locally raised revenues, from central transfers, from road funds or by cost sharing. The last section deals with planning methods and processes for investment and maintenance.

The last chapter explains the institutional implications of community roads and paths. The paper argues that it is possible to manage this network through the communities. In both Sweden and Finland the majority of roads are private, under the direct management of the users. Private ownership has proven to improve maintenance and reduce costs. Community ownership requires the setting up of cost sharing arrangements for maintenance as well as technical and managerial advice and training.

MASON, MELODY AND SYDNEY THRISCUTT (1989)
'Road deterioration in Sub-Saharan Africa'
PTRC Transport and planning annual meeting.
Brighton; pp. 19-42.
Available from: ILO/ASIST Nairobi

This study, undertaken under the SSATP Road Maintenance Initiative of the World Bank, analyses road deterioration in SSAfrica and establishes scenarios of how road improvements can be financed. Even though the study takes a conventional view on roads, it is worth reading, because it draws up catastrophic scenarios of future transport in rural areas.

Keywords:
roads,
maintenance,
SSAfrica,
financing,
institutions

The first chapter describes the bad condition of SSAfrican road networks, and the second analyses past maintenance performance. Special emphasis is laid on the (inadequate) provision of maintenance funds.

The authors state that a continuation of the existing maintenance practices would leave all unpaved roads and 80-90Êper cent of the paved priority network in a bad condition –a strategy, which is clearly not desirable. On the other hand, the rehabilitation and maintenance of the whole SSAfrican road network is not affordable. The authors push for the concentration on an 'economically justified' priority road network. They emphasize that the remaining (rural) network would stay in fair or poor condition due to financial constraints. The strategy can be financed by using annually 0.8 per cent of GNP and concentrating on rehabilitation instead of new construction. The authors emphasize that without external aid, one third of African countries is not able to restore even their priority network to an adequate condition.

The scenarios do not take into account the construction of low cost infrastructure designed for IMT and the effects of a least-cost planning approach (Howe 1996).

MORRIS, DUNCAN (1993)
'Thinking things through'
Paper presented at the 19th WEDC Conference, held on September 6-10 in Accra; 3p.
Available from: ILO/ASIST Nairobi

Keywords:
participation,
planning,
institutions,
labour-based
works

The author bases this very short paper about the provision of basic infrastructure on eight years experience as ILO advisor for labour-based works in Africa. He argues firstly for a limited role for government and the public sector by drawing a distinct line between public works and community works. This distinction should be based not so much on who pays for the works than who initiated them and who owns, manages, uses and maintains the assets. Therefore, community participation in public works should be essentially consultative rather than active during implementation. Instead of talking about communities participating in government projects, the government should participate in community projects.

Secondly, the author argues for a staged development, which maximizes coverage needs and minimizes costs and risks. In the first place a minimum acceptable level of low cost infrastructure is provided, which might be upgraded later if needed.

The third argument focuses on the optimal use of locally available resources, especially labour

force.

RIVERSON, JOHN, JUAN GAVIRIA AND SYDNEY THRISCUTT (1991)
Rural roads in Sub-Saharan Africa:
Lessons from World Bank experience
World Bank Technical Paper No. 141. Africa Technical
Department Series, Washington D.C.; 48p.
ISSN 0253-7494

The report was prepared under the Rural Travel and Transport Project of the Sub-Saharan Africa Transport Program (SSATP) presenting findings from 127 projects with rural road components. They have been hampered by the lack of coherent policy framework and institutional focus on planning, funding and maintenance. The following topics are stressed: planning of rural roads, their design and technology, resource mobilization, sectoral organization and institutional performance.

Keywords:
**planning,
institutions,
financing,
IMT,
roads,
labour-based
works**

The authors argue that rural roads only provide 'essential access'. Therefore the prime consideration for improvements should be reliability and durability rather than width and speed. This implies a concentration on spot surface improvements and better drainage instead of expensive upgrading. Labour-based methods must be systematically promoted. Programmes can be effectively implemented if road departments have adequate degree of autonomy and separate funding.

Given the lack of resources at the local level, rural road development will continue to require central funding. The participation of agricultural officers and local communities at the planning stage leads to better road selection and maintenance. Simple and well-established planning procedures encourage participation and resource mobilization at the local level. Most effective is a small centralized agency responsible for regional planning, which is allowed to raise its own funds and receives technical advice and matching funds from a central agency. The authors emphasize that transport services and production related on-farm-transports can be improved by the introduction of intermediate means of transport.

SCHROEDER, LARRY (1994)
Provision and production of rural roads in developing countries: A synthesis of research findings
USAID, Contact No. DHR 5446-Z-00-7033-00; 64p.
Available from: USAID

Keywords:
theory,
planning,
institutions,
financing,
assessment
methods,
labour-based
works

This paper was prepared as a part of the Decentralization: Finance and Management Project conducted by USAID, which analysed the failure of development projects in rural areas managed by central government agencies. The studies undertaken during the project covered seven years and fifteen countries.

After analysing the characteristics of roads, Schroeder analyses the actors and institutional arrangements in the 'road industry' (compare public service industry in Connerly/Schroeder 1996). He goes on to discuss the decisions about the allocation of funds for roads or other investments, which type of road should be built and what means have to be allocated for maintenance. The main problem is that these decisions are made by bureaucrats or technicians with no strong incentives to ensure that the public is served. This can be overcome by involving users in the decision process and creating incentive mechanisms to overcome principal-agent problems. An effective instrument would be the hiring of monitors whose salaries are tied to the desired outcomes.

The fourth chapter is about road production issues: private sector contracting, labour-based works, capital equipment and institutional arrangements for contracting. In his last chapter the author discusses financial issues. Firstly the role of central and local governments as well as donors are considered. Because there still will be a gap between resources available and spending needed, the additional mobilization of benefit-linked finances is favoured. User charges, local taxation and financing of maintenance by road user groups can be adequate sources of funds.

SHAH, ANWAR (1991)
Perspectives on the design of intergovernmental fiscal relations
PRE Working Paper WPS 726, The World Bank, Washington D.C.; 107p.
Available from: World Bank Bookstore

This paper concerns the structure of taxing and spending authorities and the manner in which intergovernmental transfers (between central and local governments) are shaped. Even though it is not directly concerned with transport infrastructure, it is worth studying, because fiscal structures are of fundamental importance in the provision of public services. However, the language used restricts the readership to those with at least a basic knowledge in fiscal economics.

Section 1 reviews theory and practice of tax and expenditure assignment and reveals that problems arise not from *de-jure* responsibilities, but from the *de-facto* administrative practice. These problems can be solved by fine-tuning of administrative procedures rather than by constitutional amendments.

Section 2 is concerned with the theory and practice of intergovernmental transfers in a federal system. Even though many countries have strong inter-regional disparities, the governments have not adopted simple programmes to equalize them. On top of that, many central government's grant programmes have no clearly-defined objectives, lack transparency, objectivity and accountability.

Sections 3 and 4 provide 16 country examples of fiscal imbalances and arrangements in the North and South. Two tables give an overview of federal/central revenue sharing mechanisms and of federal/central transfers to lower levels of government. None of the countries pay special attention to fiscal capacity or revenue potential of local governments. Allocation of funds is usually ad hoc –negating transparency, predictability and autonomy objectives. Further, the administrative capacities to closely monitor the finances of municipalities are missing. On the other hand are local governments not allowed to borrow in the credit markets and therefore relying exclusively on higher level capital transfers for investments.

Keywords: **financing, institutions, decentralization**

7. TRANSPORT AND DEVELOPMENT

AHMED, FARHAD, STEVE CARAPETIS AND M. TAYLOR (1995)
'Rural transport in Bangladesh: Impact of non-motorized transport on household's activity patterns'
Paper presented at the International Conference of the Eastern Asia Society of Transportation Studies (EASTS), Manila, September 28-29; 17p.
Available from: ILO/ASIST Nairobi

Keywords:
**Bangladesh,
NMT,
socio-economic
impacts,
household
survey,
time budget,
planning,
gender**

This short paper examines the impacts of non-motorized means of transport (NMT) on rural households in Bangladesh and examines the suitability of the Household Activity Travel Simulator (HATS) in Developing Countries.

The research team used the HATS approach to evaluate the transport activities of all members of 100 rural households in four villages. The results were summarized in a table representing the average time used for all activities (including non-transport) during the whole day. The households were than asked how their activities would change if the household was provided with a bicycle. Less than 8 per cent of the interviewees stated changes of their activity patterns. Their travel time would be reduced by 10 per cent. They would use 44Êper cent of the saved time to increase their working time and only 18Êper cent for social activities or leisure. 27Êper cent would be used for additional domestic activities, which cannot otherwise be carried out due to time constraints. The more the households were engaged in commercial activities, the more time they would spend on work.

The study found out that NMT have an enormous potential in addressing transport related problems in Bangladesh. However the impact of NMT are not as straight forward as commonly perceived. NMT are more likely to favour males and wealthier classes. The authors tested a modified HATS-approach which was found useful to evaluate transport activities of rural households in developing countries.

AIREY, TONEY (1992)
Transport as a factor and constraint in agricultural production
World Bank SSATP, IT Transport (U.K.) and ILO, Geneva; 100p.
Available from: ILO/POLDEV Geneva and World Bank SSATP

As a part of the RTT Programme of the World Bank, five village level travel and transport surveys (VLTTS) were conducted in Burkina Faso, Zambia and Uganda. The results are published as World Bank Discussion Paper No 344 (compare Barwell 1996). This unpublished report tries to assess the role of transport as a factor and constraint in agricultural production and assesses the effects of transport interventions to alleviate these constraints. Conclusions were drawn by comparing groups of households which are classified into successful, typical and unsuccessful, using agricultural income as an indicator.

Keywords:
Burkina Faso, Zambia, Uganda, IMT, time-budget, economic effects, household survey, agricultural production

For the alleviation of on-farm transport constraints, Airey emphasizes the role of IMT, which:

- shorten the time required for trips to the fields
- increase the efficiency with which loads are carried
- reduce the effort and drudgery involved in human porterage
- reduce the pest damage and spoilage at crop harvest time.

In economic terms the benefits of IMT can be considered as releasing latent factors of production, principally land, and increasing the efficiency with which the existing labour endowment is utilized. IMT enable the household to extend the distance over which agriculture is practised and they release the household's time requirements, which can be used for productive activities. The households are able to expand their agricultural production by bringing more plots under cultivation.

For off-farm transport the role of road access combined with transport services and the location of facilities are crucial. IMT play an important role in improving the speed and efficiency with which households overcome distance and transport the increased volume and weight of crops associated with a higher level of production. Airey emphasizes the positive role which bicycles may have in off-farm transport, while animal carts are primarily used for production-related transports.

The study is accompanied by many tables documented in the annex.

BARWELL, IAN, GEOFF EDMONDS, JOHN HOWE AND JAN DE VEEN (1985)
Rural transport in developing countries
IT Publications, London; 145p.
ISBN 0 94668 80 X

Keywords:
household survey, Malaysia, India, Western Samoa, Philippines, Korea, Tanzania, Bangladesh, planning

This volume contains a selection of case studies which highlight transport patterns, means and policies. The book has two related purposes: first, to improve understanding of the nature of the transport needs of rural people in developing countries and of the extent and means whereby they are currently met and second, to contribute to the development of practical policies to provide transport facilities which will better meet the needs of rural communities. The core of the book is a series of case studies covering nine developing countries in Africa, Asia and the Pacific. They are organised into three parts. The first covers the transport patterns of rural communities in Malaysia, India, Northern Nigeria and two Kenyan villages. The second looks at particular means of local-level transport in Western Samoa, the Republic of Korea and the Philippines. The studies that form the third part are more concerned with evaluations of transport policy and how it affects rural communities in Tanzania, Bangladesh and small farmers in Kenya. The common, and distinctive, feature which links the case studies, however, is that they examine transport conditions and problems from the viewpoint of rural people –small farmers, village households, etc.– by focusing attention on the nature of small farmer and household transport needs and on the physical and other constraints within which these must be satisfied. The studies pay explicit attention to the transport activities that take place remote from the motorable road network.

While all studies illustrate the local transport system in some manner, they did not have a common research framework.

- The study on Malaysia focused on the transport requirements of smallholder agriculture in a particular sub-district of the State and was carried out in 1981. It also provided some ideas on how that demand could be met.
- The study on India is based on the findings of a 1977 nationwide socio-economic study by the Ministry of Shipping and Transport on animal cart transport. It is supplemented by a survey of 200 households and 30 cart manufacturers carried out in 1980.
- Nigeria has one of the highest road densities in Africa. The case study from this country examines the transport demands of the rural population in one area of Northern Nigeria and the ways in which these demands were met, as well as attempting to evaluate the likely response of farmers to improvements in transport. It was carried out in 1981.
- The study which forms the basis for the chapter on two Kenyan villages was carried out in 1980. It was an attempt to compare the effect of contrasting levels of access on two villages in proximity to each other, focusing on the level of vehicle ownership and on the methods and costs of moving farm produce to market and meeting other transport needs.
- The study on which the chapter on Western Samoa is based was commissioned in 1980 to investigate the most appropriate rural transport technology to meet village agricultural needs.
- The chapter on the Republic of Korea describes the work carried out at Soong Jun University to improve the efficiency of the chee-geh, a traditional load-carrying frame used in rural areas.
- The objective of the study on the Philippines, carried out in 1982, was to present a description of the transport situation as it pertains to the majority of the population in a particular area of the coun-

try. It compares the most common forms of transport: jeepneys, trimobiles, railway skates and carabao sleds.
- The study of Ukinga area of Tanzania, carried out in 1981, was concerned with developing an approach to transport based on a qualitative assessment of the relationship between transport and development.
- Rural transport facilities in Bangladesh are extremely poorly developed. The study on this country was initiated by the Government to investigate the present and likely future needs for rural transport, with particular focus on the role and consequences of rural transport improvements in relation to rural development. The study was carried out between 1975 and 1977.
- The final case study, on Kenya, is concerned with general transport policy and how it affects rural farmers. It was carried out in 1976.

The final chapter reviews the most important findings of the case studies, draws on supporting evidence from other sources and discusses the implications of these findings for the development of more effective policies for the planning and provision of local level rural transport facilities. One of the main conclusions of the study is that transport planning in most developing countries takes insufficient account of the needs and requirements of the bulk of the rural population. Only the transport needs of a few large farmers producing for the export markets are captured by conventional transport planning; the transport needs of the rest of the rural population are largely for the movement of small loads over relatively short distances and this is largely ignored. In addition, poor credit facilities are a major constraint on the small producer acquiring even simple means of transport.

COOK, PETER AND CYNTHIA COOK (1990)
'Methodological review of the analysis of the impacts of rural transportation in developing countries'
in: *Transportation Research Record* No. 1274, National Research Council, Washington, pp. 167-178.
Available from: ILO/ASIST Nairobi

This paper reviews rural transport studies carried out over 20 years with emphasis on the methodologies and the underlying causal relationships. The authors argue that present impact methodologies focus narrowly on agricultural effects and fail to predict the significant increases in non-farm traffic and their related economic benefit, which are signalled by the relatively high value of travel time. They therefore do not predict benefits from personal travel, i.e. non-farm employment, service accessibility and increased mobility.

Keywords:
**theory,
socio-economic
impacts,
assessment
methods,
planning**

Next the authors set up a causal model for the assessment of rural road impacts, which defines the relationships between access change and rural economic development. The changes in transport conditions and costs and the shift in regional and village accessibility are assessed. The model only comprises the improved accessibility to local and regional markets, but does not include access to village facilities like energy and water supply. The authors propose to combine models which forecast:

- primary production by using a two stage Cobb-Douglas model
- local manufacturing and consumption by using a multiplier model
- personal mobility by using employment predicting models for work related trips and a gravity model for non-working trips
- migration based on labour supply and demand
- demand for services related on accessibility combined with a model of accessibility change.

EDMONDS, GEOFF AND COLIN RELF (1987)
Transport and development
A Discussion Paper, ILO, Geneva; 37p.
Available from: ILO/POLDEV and ILO ASIST NAIROBI

Keywords:
planning,
theory,
assessment
methods

This paper provides a description of the analysis that has been the foundation of the ILO's work on rural transportation in Developing Countries.

The first section of the paper is a broad and descriptive overview of the role of transport in development. Historically, transport has been seen as the prime causal factor in the development process. More recently, the relationship between transport and development has been recognized as more complex. Contrasts are drawn between the historical development of transport in the more developed countries and the incompleteness and polarization of transport systems in the developing countries. Efforts and investments to improve the transport system in rural areas have been directed at overcoming deficiencies, redressing imbalances, reducing economic distortions and 'catching up' rather than as a complement to growth or as a result of confident predictions.

The core of the paper's analysis is contained in the section on critical issues and limitations in transport planning. The authors argue that the assumption of a causal relationship between the provision of transport infrastructure and economic development resulted in transport expenditure being perceived as 'hard' investment amenable to quantified predictions of impact (in particular the generation of additional agricultural production) and to rigorous economic appraisal. Standard methods of economic appraisal have been demanded to protect the interests of borrowing countries and to preserve confidence in international capital markets. Changes in development planning and increasing emphasis on investments on rural roads led to efforts to redefine methodologies for the economic appraisal.

The authors of this discussion paper suggest that alternative methodologies may go further to reveal the total potential benefits of a given investment. They stress the importance of non-economic considerations in planning and the need to take account of other non-quantifiable social benefits in determining the viability of specific investments.

The authors' examination of equity considerations and the distribution of benefits from investment in transport leads to a nuanced appreciation of the links between infrastructure and the development process that goes far beyond the earlier assumption of a direct causality. More investment in transport infrastructure and means is necessary, but it is unrealistic to expect that more of the same types of investments would make a significant impact on rural poverty. In order to make a significant impact, the authors suggest that:

- The physical design of transport infrastructure should be appropriate to the means of transport and types of transport services that rural communities need and can afford.
- Complementary policies and measures are required to create the physical, institutional and economic environments so that the poor can derive more benefits and so that the development effects anticipated will actually be created.
- Transport must be considered in a broader context. The use and impact of rural road networks once they are constructed are as important as their development. Rural communities should be involved in the planning of transport investments.

Finally, the authors argue that conceptual refinements are considerably further advanced than practical applications and that there remain a number of obstacles to innovation. There is still uncertainty about how, practically, to design improvements in access to appropriate transport services in rural areas. Entrenched economic appraisal and planning procedures restrict innovation. Despite these constraints, the authors point out that pilot projects and structural adjustment loans might be two ways that new ideas about transport planning could be translated into action.

HINE, J.L. ET AL (1983)
'Accessibility and agricultural development in the Ashanti Region of Ghana'
in: *TRRL Supplementary Report* 791, Transport Research Laboratory, Crowthorne, Berkshire; 33p.
ISSN 0305-1315
Available from: TRL

HINE, J.L., J.D.N. RIVERSON AND E.A. KWAKYE (1983)
'Accessibility, transport costs and food marketing in the Ashanti Region of Ghana'
in: *TRRL Supplementary Report 809,* Transport Research Laboratory, Crowthorne, Berkshire; 25p.
ISSN 0305-1315
Available from: TRL

Keywords:
Ghana,
economic effects,
accessibility,
marketing,
roads,
transport costs

These two reports reflect the 'historic' works of the Transport and Road Research Laboratory (TRRL, nowadays TRL) in the Ashanti region of Ghana. The research collected data from 491 farmers in 33 villages in the cocoa region 8 to 100km away from the regional capital Kumasi.

The first report researches the effects of road accessibility on marketing of agricultural produce. The authors found little evidence that a direct nexus between motorized accessibility and marketing exists, because the difference in transport costs were marginal due to the small distances. Stronger evidence was found regarding non-agricultural activities, which was stronger close to the regional capital.

The second report observes transport costs and their effects on food marketing. Transport charges accounted for only a small proportion of the wide difference of food market prices. Therefore the improvement of road surfaces had negligible effects on producer prices and thus on marketing. However the replacements of footpaths by vehicle tracks have 100 times stronger effects than upgrading the same length of roads to good quality surface.

SIEBER, NIKLAS (1996)
Rural transport and regional development: the case of the Makete District, Tanzania
Karlsruhe Papers in Economic Policy Research, Vol. 4, Nomos Verlag, Baden-Baden; 190p.
ISBN 3 7890 4507 1

This PhD thesis examines the nexus between appropriate transport interventions and rural economic development.

The first appraises the economic impacts of the Makete Integrated Rural Transport Project conducted by the ILO in Tanzania. Footpaths, tracks and roads were improved, Intermediate Means of Transport (IMT) promoted and transport-avoiding measures implemented. The author surveyed 170 household and compared the transport patterns to past research which had been conducted before the project started. The economic benefits of different appropriate transport improvements are assessed and compared on the household level.

The study comes to the conclusion that comparing the absolute effects and the cost efficiency it can be safely stated that non-motorized transport interventions have the same magnitude of impact as interventions in the motorized sector.

In the next part of the thesis the author establishes an econometric model, which simulates the observed economic impacts by using a systems dynamics approach. Six scenarios describe the effects of various transport interventions on the regional economy. The scenarios include footpath-, track- and feeder road-construction, a credit scheme for the purchase of IMT and the financing of a rural road by user charges.

In the last part, current methodologies and practices in rural transport investment appraisals are analysed. The author proposes a broader approach including non motorized transport investments by incorporating a household survey into cost/benefit assessments. The addition of monetarized time savings to total benefits leads to a more objective measure to assess non-motorized as well as conventional transport interventions.

Keywords:
Tanzania, economic effects, IMT, assessment methods, planning, household survey, roads, time-budget

HOWE, JOHN AND PETER RICHARDS (EDS) (1984)
Rural roads and poverty alleviation
Intermediate Technology Publications, London; 168p.
ISBN 0 946688 05 2

Keywords:
Egypt,
India,
Botswana,
Thailand,
poverty alleviation,
roads,
socio-economic impacts,
assessment methods,
planning

This book is a collection of seven papers on various aspects of the question of the relationship between rural roads and poverty alleviation. The authors provide a concluding chapter which discusses the main conceptual points raised in many of the papers and attempts to propose possible future directions for action.

Chapter I is an essay on the economic context of rural roads. It looks at the economic function of transport and mobility, roads and agricultural modernization and personal mobility. The chapter also includes an example of the analysis of road programmes in Nepal. The author of this chapter argues that if poverty alleviation is to be one of the principal aims of rural development policies, road selection criteria should be redesigned to meet this objective. The use of social cost-benefit analysis may not solve the problem of targeting benefits to the poorest segments of society. What is necessary is an understanding of the interaction of road construction and improvement with other socio-economic factors in rural development.

Chapter II reviews the various selection criteria which have been used in developing countries for road investment programmes, and looks in particular at the extent to which these criteria have managed to take into account the income distribution implications of road programmes, through explicit or implicit assumptions of the means by which socio-economic groups are expected to benefit from the road investment.

Chapter III reviews available evidence from impact studies on the effects of road investment programmes on income distribution and access to services, and discusses where relevant the selection criteria used and the other assumptions made in planning the road programme in question. The paper focuses in particular on the mechanism by which various socio-economic groups were expected to, and did, benefit from the programmes or by which an expected flow of benefits was obstructed. The review reveals the paucity of evidence on the effects of road investment programmes on rural incomes and income distribution.

Chapters IV, V, VI, VII are case studies of rural roads and poverty in Egypt, India, Botswana and Thailand. In Chapter VIII, the editors attempt to draw conclusions from the preceding chapters. They reiterate the main idea underlying the book: if income distribution and poverty alleviation are to be the main aims of rural development policies, then the criteria used for deciding road investment and improvement may have to be redesigned. In most cases up to the mid-1980s, an area's potential contribution to agricultural production and output was the sole explicit or implicit factor considered in the process of road selection. Thus, road investments could be expected to reinforce the prevailing social and economic structures and even increase stratification. Furthermore, the question of technological choice was also usually left out (with the exception of Botswana out of the four case studies). The authors point out that the provision of a road is often taken as synonymous with the provision of transport, and even studies that begin by acknowledging the need for better *transport* often finish by recommending the need for *road improvements*. However, explicit consideration of the means of personal and goods transport is essential in making transport investments, as is the examination of the conditions and priorities for road maintenance.

Chapter VIII also draws out several variations in the selection criteria by which investments have been decided. Whereas quantitative analysis is usually used or required by lending agencies, programmes initiated by developing countries tend to emphasize political or administrative judgements in deciding on investments. One of the most robust results to emerge from the impact studies is that while the most likely form of traffic on a new road is transport of people rather than goods, this has been virtually ignored in appraisal methods which continue to value only the movement of goods. Finally, the authors point out the paradoxical situation of the increasing expense of ever more complex appraisal methods with respect to the steadily decreasing cost of constructing rural roads to serve smaller and smaller communities. They suggest that more efficient and effective selection criteria would include, first, a more rigorous analysis from the geographical and demographic viewpoints in order to secure a fairer distribution of resources to poor areas and, second, attempts to persuade decision makers to view the movement needs of the poor as transport problems (vehicle and track) rather than problems of access that can be solved by the provision of roads.

SIMON, DAVID (1996)
Transport and development in the Third World
Routledge Introductions to Development, London; 194p.
ISBN 0 415 11905 7

Keywords:
**theory,
NMT,
socio-economic
impacts,
public
transport,
privatisation,
deregulation**

This book is a sound introduction for students of transport geography. It gives a broader perspective than conventional textbooks, because it considers all modes of transport, including non-motorized conveyances. The book contains twelve practical case studies and many tables, graphs and pictures which illustrate the text. A list of review questions helps lecturers to formulate their curricula.

After describing the principal transport trends, the author places different approaches to transport in the context of general development theories. He then introduces the reader to rural transport problems. By quoting examples of the negative economic and social effects of road projects, he questions the common assumption that roads generate positive development. Simon widens the reader's perspective by describing the benefits of non-motorized and low cost means of transport in rural areas. The advantages of these conveyances in the urban environment are further emphasized in the following chapter, which also contains a large section about public transport. Chapters about long distance transport and a summary with a critique of privatization and deregulation policies follow.

The author attempts to discuss a complex matter in only 190 pages. Many topics are only touched upon and others, like transport models, appraisal methodologies, financial constraints or the important issues of infrastructure maintenance are missing completely. The book gives a good overview of transport problems in the South from the perspective of a geographer, but it does not present the analytical tools to solve them.

WILSON, GEORGE W. (1973)
'Towards a theory of transportation and development'
in: B. S. Hoyle, *Transport and Development*, London; 23p.,
ISBN 00333144783
Available from: ILO/ASIST Nairobi

Keywords: theory, economic impacts, roads

Although Wilson's article is more than 25 years old and focuses only on motorized transport, it still provides a critical analysis of the economic impacts of transport interventions in developing countries. The author argues that transport investments are a 'necessary but not sufficient' precondition for development, which might as well entail negative economic effects.

Wilson analyses the economic effects of transport investments in Developing Countries. He distinguishes between (1) the creation of economic opportunity and (2) the response to economic opportunity. The first depends upon the quality and quantity of resources in the regions served, the actual change in transport rates and service and commodity price levels. Various studies show that when producer prices were falling and yields not increasing, economic development could be stimulated by declining transport costs. In a situation with rising prices and increasing production, transport investments only provided further stimulus, and were permissive rather than causal. The main factors influencing the response to new transport capacity are: (a) awareness of its potential, (b) the availability of finance, and (c) the magnitude of possible benefits relative to alternative investment options. The author states that next to the direct economic impacts, spillover effects occur, which are much stronger than the direct reductions in user costs: 'The unlimited access of roads in the early stages of development of any region has an awareness effect that serves to induce a larger number of people to take advantage of new economic potential'. These changes in attitude are much more strongly influenced by roads than by other modes of transport (not considering NMT).

Wilson mentions as well the negative impacts of transport interventions: regional markets which had been protected by high transport costs, would be exposed to competition with cheap international products. This process could even entail a decline in real incomes.

8. GENDER AND TRANSPORT

BRYCESON, DEBORAH FAHY AND JOHN HOWE (1993)
Rural household transport in Africa: Reducing the burden on women?
**Africa Studies Centre Working Paper Vol. 15,
Leiden, The Netherlands.**
Available from: ILO ASIST

Keywords:
**gender,
domestic
transport,
SSAfrica,
household
survey**

This paper briefly reviews findings on rural household transport demand in surveys and literature on East and West Africa. The findings indicate low levels of transport ownership and significant time and energy required for transport activities. Evidence of the extent of women's participation in rural transport and attitudes towards their transport role are discussed and the question raised as to why human porterage in Africa is so restrictively women's work. The impact of demographic and economic change, and inter-relationship of agriculture in determining household transport requirements and the allocation of intra-household transport tasks are considered. General approaches to alleviate rural transport constraints are outlined, focusing particularly on the constraints faced by women in accessing improved means of transport and men's tendency to appropriate the modes of transport. Finally, the authors argue that certain key aspects of women's transport work have not yet been taken into account by current household transport demand methodologies. While rural transport development has tended to assume that external interventions such as infrastructure improvements or introduction of improved means of transport are key to alleviating transport problems, households' internal management of transport and scope for improvements as a result of internal organizational changes have been overlooked. Women's multiplicity of tasks, child care role and traditional carrying strategies need to be given greater attention. The most important stumbling block to improvements in rural household transport is attitudinal. The authors summarize that the development of transport programmes which are of value to African rural women transporters depend on measures determined from the perspective of the women themselves, and a key challenge is to change attitudes that regard women as the natural load carriers.

Curtis, Val (1986)
Women and the transport of water
IT Publications, London, U.K.; 48p.
ISBN 0 946688 42 7

One of the most arduous and time-consuming daily tasks of rural women in developing countries can be the haulage of water from often distant sources. This illustrated paper looks at some of the problems women have with this work and investigates whether improved means of transport could help relieve their burden. The paper is divided into two parts. The first looks at the advantages of improving water supply, and the health and economic effects of carrying water. It introduces some alternative transport ideas which range from carrying aids to animal carts. The second part, based on field research in Kenya, briefly reviews the nationa water supply situation, the water transport workload faced and a range of carrying methods used in different parts of Kenya. It also provides three case studies of project communities in poorer, drier areas of Kenya with transport and water supply problems and suggests ways in which improved transport could help in the three selected communities, particularly through using donkeys.

Keywords:
**gender,
water,
IMT,
Kenya,
domestic
transport,
donkeys**

DORAN, JO (1996)
'An imbalanced load: Gender issues in rural transport work'
Draft paper prepared for ITDG, U.K.
Available from: ILO/ASIST Nairobi

Keywords:
gender,
time-budget,
domestic
transport,
social impacts

This paper seeks to provide some insights into the gender issues involved in rural transport and on implementing gender-responsive transport programmes and activities. The paper briefly outlines the current general state of both gender and rural transport analysis before presenting reasons for adopting a more gender equitable approach in rural transport. A number of gender issues for consideration in transport work are then discussed including transport roles, access to improved transport, benefits from improved transport, child care and reducing gender disparities in transport. A framework is outlined for analysing the gender gap followed by some lessons from experience. The author concludes that the time and effort that rural women spend on transport activities, particularly domestic activities of water and firewood collection, is an impediment to their using time more productively in other activities. The rigidity in the gender division of labour when carrying is done by head, back or shoulder loading and women's limited access to improved means of transport are seen as key determinants in the current imbalance of transport workload. Improving their access to more efficient means of transport with a more equitable household division of tasks, and improving rural infrastructure such as access to closer water supplies can make substantial contributions to reducing the gender gap in transport activities. Transport interventions need to take more account of women's perspectives, their existing transport technologies and their child care responsibilities besides involving women more centrally in the planning and implementation of such interventions.

DORAN, JO (1996)
Rural transport. Energy and environment technology source book
UNIFEM, Intermediate Technology Publications, London.
ISBN 1 85339 345 2

This source book seeks to raise awareness and to provide information on how rural transport problems might be identified and addressed. Since much of rural transporting work is done by women, efforts to address transport needs must obviously pay attention to women who carry the main burden. The book starts by highlighting rural transport activities and needs using data from southern, eastern and western Africa. Women's workload and transport burden in their daily domestic activities are considerable. Chapter 2 focuses on identifying transport needs. Information on ways of addressing those needs with different means of transport, improving local transport infrastructure and establishing transport services is given in Chapter 3. Also included in that chapter is a brief discussion of non-transport interventions such as improving water supplies and increasing the availability of fuelwood, grinding mills and other infrastructure, all of which can substantially reduce the time and effort spent on transportation. There are constraints which are frequently encountered in trying to assist women and their communities to gain access to improved means of transport. Some of these are discussed in Chapter 4. The case studies in Chapter 5 illustrate approaches and technologies which have been tried and some of the conclusions to be drawn from the experiences. Chapter 6 provides a checklist of questions for consideration before identifying or acquiring a means of transport. A number of references and contacts are given in the final chapter.

Keywords:
gender, environment, IMT, domestic transport

HOWE, JOHN AND DEBORAH FAHY BRYCESON (1993)
'Women and labour-based roadworks in Sub-Saharan Africa' in: *Labour-based technology: A review of current practice. Proceedings of the regional seminar on labour-based road technology, held on September 27-October 1 in Harare, Zimbabwe, organized by the Institute of Engineers.* ILO, Geneva; 16p.
Available from: ILO/ASIST Nairobi

Keywords:
gender, labour-based works, SSAfrica, Kenya

Reviewing project-related literature, this paper explores the incidence of, and attitudes towards, rural women's paid participation in labour-based road works. Synthesizing findings regarding the social and economic impact on female participants and their households suggests that labour-based road works meet an essential employment need for some categories of women. The paper raises some theoretical considerations before briefly reviewing project experience, particularly that of the Kenya Rural Access Roads Programme. The review covers experiences on attitudes towards the suitability of roadwork for women, including that of men, and states that the availability of alternative sources of employment and the relative wages appear to be the main determining factors. Recruitment and participation rates are reviewed, as well as the household implications of women's participation. The authors also look at the question of whether all women benefit from participation on labour-based roadworks. Based on the projects' varied experience, the authors find it extremely difficult to come to any general conclusions on the benefits and disadvantages women experience when participating in labour-based rural roadworks. Nonetheless, there is abundant evidence that labour-based roadworks are a welcome source of income for asset-poor women. Furthermore, the demonstrator effect of women undertaking non-traditional roadwork tasks could help to challenge fixed notions of current gender roles.

JENNINGS, MARY (1992)
Study on the constraints to women's use of transport in Makete District, Tanzania
ILO, Geneva; 30p.
Available from: ILO, Geneva

This report is based on a consultancy undertaken during the Makete Integrated Rural Transport Project's consolidation phase. The project is located in a remote, rural and hilly part of Tanzania. The report identifies constraints to women's use of means of transport in Makete District and makes recommendations on how the economic constraints can be overcome. Under socio-cultural constraints the study found a wide discrepancy in the capacity of households to invest labour and financial resources in, and benefit from transport related interventions. The initial project data collected and analysis of transport problems at the household and intra-household level was not translated into planning and implementation, thereby limiting the positive impact of the project on women from the outset. In the study, female headed households were found to be significantly poorer compared with the minority of households headed by men, and used fewer project-introduced wheelbarrows or donkeys. It was questionable whether the project had made any significant impact on household transport demands. Under economic constraints, the study suggested that women are unable to benefit from the project interventions because either the opportunity cost to women of investing in means of transport was too high in a context of struggling to provide for basic needs, or, that the self-help path and track activities which were predominantly implemented using women's labour had placed an additional burden on women and reduced their time spent on subsistence agriculture. Technical constraints included questioning the viability of wheelbarrows in hilly terrain for carrying water and needing two to load donkeys when women tend to work alone. The study findings suggest that addressing rural transport needs, and especially those of women, can only be resolved within the context of tackling poverty, and enabling women to meet their basic needs. In Makete, assisting women to increase their subsistence agricultural output might be likely to achieve this. In summary, the report uses the project's experience to raise the critical question of

Keywords:
**gender,
IMT,
poverty,
Tanzania,
planning**

whether problems which manifest themselves as transport problems are indeed so, or whether the solution is to be found by addressing rural infrastructure problems of which transport is but one component.

MALMBERG-CALVO, CHRISTINA (1994)
Case study on the role of women in rural transport: Access of women to domestic facilities
SSATP Working Paper 11, World Bank and Economic Commission for Africa, Washington, D.C.; 59p.
Available from: World Bank SSATP

Keywords:
gender,
accessibility,
domestic
transport,
water,
firewood,
SSAfrica

This report, reviewing available documentation on access of rural women in Sub-Saharan Africa to domestic facilities, investigates the magnitude of women's domestic transport burden and aims to assess the impact of 'non-transport interventions' on women's transport work. 'Non-transport interventions' include installation of improved water supplies, development of community woodlots, promotion of more efficient wood-burning stoves, and establishment of crop grinding mills. Chapter 2 analyses the existing travel patterns of rural households in four study areas (Ghana, Zambia and two in Tanzania), and quantifies the time and effort spent on transport by activity and gender. The study confirms that in Sub-Saharan Africa it is women, assisted by daughters, who are predominantly responsible for the tasks of water and firewood collection and travel to the grinding mill and that this work is both burdensome and time-consuming. Chapter 3 studies the experiences from projects concerned with provision of domestic and subsistence facilities and their effect on women's time and energy. It is clear from documented experience that the potential to lessen women's transport burden is often not realized. A common theme underlying many of the projects having only limited impact is the lack of women's involvement in their design and implementation. Chapter 4 focuses on ways to stimulate rural women's participation in programmes to alleviate their domestic burden. It concludes that, if such projects are to have positive impact, their design must be based on an understanding of the local situation of women; must incorporate the expertise, knowledge and perceptions of women and must substantially involve them in the project's implementation.

MASCARENHAS, O. (1995)
Guidelines for incorporating gender issues in rural transport planning
ILO, SDC, Dar Es Salaam; 51p.
Available from: ILO/ASIST Nairobi

This consultancy report is a part of the guidelines on rural transport planning produced by the ILO for Tanzania.

Part I gives a brief overview on gender issues in rural transport, while Part II provides practical guidelines how to incorporate gender issues into transport plans in six steps. The first step outlines the need to incorporate gender issues in the conceptual framework of rural transport plans. Step II comprises the collection of data and the establishment of accessibility data base disaggregated for men and women. During step III a ranking of access problems is undertaken before, in step IV, strategies are developed to address these problems. The last two steps prioritize locations for interventions and integrate them into (regional) development plans. The last chapter is concerned with institutional issues and monitoring and evaluation. The strategy does not consider the costs of construction and maintenance.

Keywords: gender, planning, institutions, household survey

McCALL, M.X., (1985)
'The significance of distance constraints in peasant farming systems with special reference to Sub-Saharan Africa'
Applied Geography 5: 325-345.
Available from: ILO/ASIST Nairobi

Keywords:
agricultural production, accessibility, gender, resettlement, SSAfrica, economic effects, time-budget

Analysis of agricultural development potential at village level tends to neglect the factor of relative location, compared with the attention paid to physical resources and economic factors. This paper argues that, in African peasant agriculture, distance takes on increasing significance when farming populations are resettled and agglomerated. The impact of agglomeration and excessive 'journeys to work' are identified as affecting the quantity and the quality of agricultural labour inputs, the collection of domestic necessities (especially fuelwood), livestock husbandry, and socio-cultural and welfare conditions. Some simple analyses of time-distance relations, such as the 'effective working day' are also described, and a model of peasant decision-making with respect to optimizing farm activity location is proposed as a descriptive-explanatory tool. Response to distance problems is considered as part of rural change; and the particular position of peasant women vis-à-vis distance and transport technology is stressed. Data collection methods and descriptive statements of the spatial relations within a village, or an agro-ecological zone, are outlined within the framework of rapid rural appraisal. Finally, a number of potential solutions to the agro-economic distance problem are briefly discussed – either as changes in farming systems, or as redistributions of the working population. Changes with the greatest potential are intensification and satellite settlements, though both face difficulties in policy and in implementation.

OPIYO, R. (1995)
Women, water and firewood fetching: Reducing the burden and raising the income
IT Transport Consultancy commissioned by the ILO as part of the Provision of social infrastructure and women's use of time project, ILO, Geneva; 84p.
Available from: ILO/ASIST Nairobi

This consultancy report combines a brief review of homestead work and water supply and agroforestry programmes with the findings from two Kenyan projects improving water supply and fuelwood availability. The report starts by outlining categories of homestead work and factors contributing to variations in homesteads and women's transport burdens as well as time spent on water and firewood collection. It then briefly looks at ways of reducing the burden through improving water and firewood supplies and measuring the effects and benefits. The first case study from Western Kenya relates to a tubewell water supply project. Findings cover water collection patterns pre- and post-project, perceived benefits of the improved water supply, reallocation of time saved and use of extra water particularly for economic activities. Community ideas on increasing the effects and benefits of the improved water supply project are given. The second case study relates to a water tank and agroforestry project also in Western Kenya. Time savings were much less apparent in this case. The author suggests that more attention should be paid when implementing projects aimed at reducing women's time in water and firewood collection to identify and encourage women to engage in more productive activities with the time saved. Improved water supply and agroforestry projects which result in noticeable time saving for women in water and firewood collection, and greater use of water, often lead to both direct and indirect economic benefits for the homesteads involved.

Keywords:
**gender,
water supply,
firewood,
domestic
transport,
time-budget,
Kenya,
economic
effects**

OVERTON, K. AND A. ZAMBEZE (1997)
Case study of bicycle project in Mozambique
Part of ITDG project 'Development of guidelines for dealing with gender issues in transport work', ITDG, U.K. (Final report forthcoming)
Available from: ITDG, UK

Keywords:
gender,
credit,
bicycle,
Mozambique,
IMT,
NMT

This case study report, looking specifically at gender and transport issues, describes a project implemented by AMRU (National Association for the Development of Rural Women) and ITDP (Institute for Transportation and Development) being implemented in Chockwe, Mozambique. The project is importing used bicycles from America at very little cost, and selling them at a price which covers the bicycle costs to rural women through a credit scheme run by AMRU. The cost of the bicycles is considerably cheaper than the cost of locally available bicycles. The project is also establishing spares and maintenance support. The report describes the existing transport situation in the project area, household transport characteristics, community perceptions and attitudes on transport and the project's activities. Whilst this case study was undertaken at an early stage of the project's implementation, the report gives considerable information on rural transport in Mozambique and some of the early challenges and lessons learnt by the project. The authors conclude with useful recommendations to ensure that women benefit equitably from a transport project, although the emphasis is weighted towards bicycle projects.

RAO, NITYA (1993)
'Mobility for rural women: A cycling campaign in South India'
Article written as a case study for inclusion in UNIFEM *Rural Transport Sourcebook*
Available from: ILO/ASIST Nairobi

SAINATH, P. (1995)
'Where there is a wheel: The humble bicycle has freed over 100,000 rural women in Tamil Nadu'
Humanscape, July.
Available from: ILO/ASIST Nairobi

Both these articles describe the mobility component of the Total Literacy Campaign or *Arivoli Iyakkam* (Light of Knowledge Movement) launched in Pudukkottai District in Tamil Nadu in

1991 and the dramatic increase in mobility and self-confidence of thousands of semi-literate women as a result of learning to ride bicycles. Rao's article describes the context, the origins of the cycling campaign and its rationale, how the women learnt to ride through hiring, teaching others, cycle camps, competitions and demonstrations and describes how it was financed. Some examples of confidence and initiative acquired are given and the author concludes that literacy or education for an Indian woman does not merely imply knowledge of the three R's, but its functionality in their everyday life and the creation of awareness and self-confidence are equally important aspects. She suggests that the example of the Total Literacy Campaign throws up strategies for development, perhaps in a different field such as transport, which are worth pursuing in order to meet both the practical and strategic needs of poor women. The journalistic article by Sainath also describes the campaign having interviewed a number of women cyclists and organizers.

Keywords:
gender,
bicycle,
education,
mobility,
India,
IMT,
NMT

URASA, IRENE. (1990)
Women and rural transport: An assessment of their role in Sub-Saharan Africa
Rural Travel and Transport Project, SSATP, ILO, Geneva; 79p.
Available from: ILO/Geneva

This study paper is part of the Rural Travel and Transport Project series of policy and issues papers and draws on a wide array of data to portray women's role in society and their transport tasks as well as look at the benefits and constraints to their acquiring improved means of transport. The first part reviews women's rural activities and associated transport tasks. Domestic transport tasks comprise the higher workload, followed by provision of food. The second looks at 'intermediate technologies' for reducing women's transport tasks including transport and non-transport interventions and constraints to their impact and dissemination as well as their benefits. The author concludes that women's contributions to rural transport is considerable and that it plays a major role in their productive and domestic lives. In addition, low cost transport interventions have the potential to improve women's accessibility and personal mobility and free their time for other activities. However, the interventions are narrowly assumed to be manufactured equipment.

Keywords:
women,
domestic
transport,
SSAfrica,
IMT

9. MISCELLANEA

BARWELL, IAN (1996)
Transport and the village: Findings from African village-level travel and transport surveys and related studies
World Bank Technical Paper No 344,
Washington D.C.; 66p.
ISSN 0259-210X

Keywords:
household survey, gender, planning, IMT, transport services, planning

As a part of the RTT Programme of the World Bank five village level travel and transport surveys (VLTTS) were conducted in Burkina Faso, Zambia and Uganda (compare Airey 1992). Part I of the booklet summarizes the research findings of these studies comprising:

- household transport patterns (trips, time spent, transport volume, modal split)
- women in transport (gender distribution of transport purposes, effort for subsistence transports, cultural constraints)
- Role and economics of IMT (the role of bicycles and oxcarts for production and marketing of agricultural produce)
- rural roads and transport services (road access and household income, availability of transport services.

Part II draws conclusions on how the rural access to goods and services can be improved in SSAfrica. The problems of energy and water supply can be most effectively solved by locating the sources close to the households and the use of fuel-efficient stoves. Crop production and marketing can be improved by increased use of IMT and footpath improvements. Locational planning might influence the accessibility to social services, while IMT are favourable to increase access to non-agricultural income.

BARWELL, IAN AND CRISTINA MALMBERG CALVO (1988)
Makete Integrated Rural Transport Project: The transport demands of rural households, findings from a village-level travel survey
Rural Transport Paper No 19, ILO Geneva; 90p.
Available from: ILO/Geneva

This study observes the transport patterns of rural households in Makete District, Tanzania, which was undertaken as an ex-*ante* survey for the Makete Integrated Rural Transport Project. An ex-*post* survey (Sieber 1996) evaluates the economic impacts of this project.

Keywords:
household survey, gender, time-budget, Tanzania

The survey questioned 431 rural households in 19 villages in the Makete District about their household endowment, expenditure, transport behaviour, agricultural production and marketing. The annual transport burden of a five person household amounts to 87 tkm, with 72 per cent being carried by women. The study revealed that more than 80 per cent can be regarded as internal transport in and around the village. Transport for water and firewood collection, to the fields and to the grinding mills account for 98 per cent of the tkm. The remaining transport consists of trips to the village centre, to health facilities and to markets. An average household spends more than 2500 hours annually on transport.

The document also gives insights to rural transport survey methodologies: design of questionnaires and interviews, selection of study areas, choice of interviewed households, conduction of village leadership discussions and village mapping methods. The annex presents the questionnaire and the detailed results.

HEIERLI, URS (1993)
Environmental limits to motorization
SKAT, St. Gallen, Switzerland; 200p.
ISBN 3 908007 41 2

Keywords:
environment,
motorization,
NMT,
bicycle,
urban transport

This book gives comprehensive information on environmental effects caused by motorization in urban areas. The author argues that severe environmental problems would occur if the level of motorisation already existing in developed countries is achieved in the South as well. He proposes the increased use of electrical and non-motorized transport (NMT) to meet these problems. The author uses examples from all over the world to support his argument, which are clarified by graphs, tables and pictures. The book provides a useful overview of future transport trends and their impacts.

In the first chapter Heierli describes the wish of many people in the North and the South to own a motorized vehicle. The second chapter gives examples of how the vehicle population in the South is increasing and the effects this will have on air pollution and global warming. The author discusses the advantages and limitations of a sustainable transportation system using NMT and electric vehicles. The fourth chapter gives many examples of the 'bicycle boom' during the last decade in some European cities. The next chapter describes the transport needs in the Third World: examples are given about transport policies, constraints to NMT, their use in urban and rural areas and gender issues. The last two chapters give examples of a strategy to promote NMT by producing cheap bicycles, introducing credit schemes and adopting a different transport planning approach. Finally Heierli lists the restrictive policies in cities which should be adopted towards private cars and the promotional policies in favour of low pollution modes and public transport.

HOWE, JOHN (1996)
Transport for the poor or poor transport?: A general review of rural transport policy in developing countries with emphasis on low income countries
IHE, ILO, Delft, Geneva; 77p.
Available from: ILO/POLDEV Geneva

Howe gives a comprehensive review of rural transport policies, which might serve as an introduction to the whole subject.

In his first chapter Howe presents theories of transport investments and development: he concludes that despite a huge weight of theoretical and empirical evidence, modern transport investments (in motorized transport), are still regarded as catalytic for development. This policy is wasteful, because it concentrates mainly on the provision of roads and does not take into account the supply and appropriateness of vehicles.

The next chapter describes the historic development of transport systems in the South and criticises the current roads and car approach, which regards non-motorized transport as marginal. At the end of the 1970s non-motorized vehicles transported about 80 per cent of the tonne-kilometres in India. China is quoted as a good example of how an economy can achieve strong economic growth rates by not following the roads and cars dogma. Howe criticizes current investment appraisal methods because they favour conventional interventions.

The third chapter describes the foundations of an appropriate planning approach, which is based on user-demand studies, low cost vehicles, household travel analysis (see Barwell, 1996) and Integrated Rural Accessibility Planning (IRAP).

In his last chapter, Howe proposes a new planning vision based on IRAP combined with a least total cost planning approach (LTC). While IRAP focuses on the mobility needs of rural people, LTC tries to satisfy these demands by minimizing the use of financial resources.

Keywords:
**planning,
assessment methods,
socio-economic impacts,
IMT,
theory,
China,
India,
poverty**

THE WORLD BANK (1996)
Sustainable transport, priorities for policy reform
**Development in Practice Series, Washington D.C.;
131p. and 11p. of references**
ISBN 0 8213 3598 7

Keywords:
environment,
policy,
deregulation,
privatization

This book, published in May 1996, distills the lessons learned by the World Bank in developing and transitional economies. In preparing this book, a wide range of groups were consulted inside and outside the Bank which produced a comprehensive view of the transport sector. The term 'sustainable' is used in a broader sense; it comprises not only economical and financial, but also environmental and social sustainability.

The chapter about **economical** sustainability repeats the well-known (and justified) demands to create a 'competitive market environment within a framework of public control', to improve public sector management and to set prices to efficient levels.

The chapter concerned with **environmental** sustainability focuses on the reduction of the damaging environmental effects of motorized transport. The bank proposes to combine the decrease of transport costs with the reduction of environmental effects, for example by promoting low consumption vehicles. The introduction of pollution and congestion charges might be an economic instrument to achieve this target as well as changing the modal split towards more environmental conveyances. The reduction of transport accidents is another important topic of this chapter.

Social sustainability focuses on policies to help the poor. The issues: provision of transport services, non-motorized transport, gender, accessibility, participative planning and labour-based works, are discussed in this publication. Unfortunately the book fails to emphasize the economic benefits stemming from many appropriate transport interventions. However, the bank stresses the positive experience in co-financing rural works by municipal funds and local communities.

The bank states that there is a **changing focus in transport policy**: the supply of transport infrastructure and services will shift from governments to the market.

Therefore a strong institutional framework is required, which may regulate the market for example by setting charges that reflect true costs (including environmental and congestion costs). Economic justifications of transport projects will take into account cross-modal effects including non-motorized transport and value time savings. Community participation is regarded as a necessary complement because effective markets do not exist in many poor countries. However, a devolution of responsibilities to local governments is often problematic owing to lack of qualified staff and funds. Only 3 per cent of public employment is with local governments (industrial countries 11 per cent) which have to manage more than 80 per cent of the length of national road and track networks. Effective rural transport planning involves a decentralization of responsibilities and financial means combined with staff training in participatory planning methods.

10. PROMOTIONAL MATERIAL

ALI-NEJADFARD, FATEMEH AND J. BIJL (1995)
Integrated rural accessibility planning and access interventions
Information Kit, ILO/ASIST, Harare
Available from: ILO/ASIST Nairobi

Keywords:
**IMT,
NMT,
design,
manufacture,
technology,
low-cost
vehicles**

This folder contains 10 loose leafs and a poster about Integrated Rural Accessibility Planning (IRAP). The first set of papers describes briefly the basic concepts of IRAP. Two papers list ongoing IRAP projects and recommended literature. The remaining two papers contain graphs about the access needs of rural households and rural travel and transport patterns. The A2 poster is about community participation in transport planning and rural development. The document focuses mainly on the social needs of rural households and less on the economic effects of appropriate transport interventions.

ILO (1994)
Rural travel and transport, Makete, Tanzania: a case study. Introduction to a new approach to rural accessibility and transport
Video VHS-PAL, ILO/POL-DEV, Geneva, 15 min.
ISBN 92 2 109479 0
Available from: ILO/Geneva

The video outlines the new approach towards rural travel and transport. It reports from the Makete Integrated Rural Transport Project in Tanzania where footpaths, tracks and roads had been improved, IMT introduced and facilities relocated. The video is not a film but rather a slide-show on TV. It contains not only pictures about transport in Makete, but also graphs clarifying the text and slides about IMT, improved footpaths and labour-based works.

ILO/ASIST (1996)
Responding to the challenges
Development Engineering, Nairobi, Kenya; A2 poster
Available from: ILO/ASIST Nairobi

The front-page of the poster depicts transport problems on the left hand side and the improved situation on the right, while appropriate transport interventions are drawn in the middle of the poster. The back gives detailed information on ILO ASIST NAIROBI.

ILO/ASIST (1995)
Development engineering poster series
Harare, Zimbabwe; 6 posters A2 size
Available from: ILO/ASIST Nairobi

A series of six posters which illustrates some of the benefits of the use of labour-based technology for road works;
- *Economic benefits* – depicts employment, local investment, development of local manufacturers derived from using local resources as opposed to the importation of foreign resources.
- *Equipment choice* – an illustration comparing the costs of using foreign equipment versus intermediate locally manufactured equipment to carry out the same roadworks.
- *Machine or jobs* – an illustration comparing the costs of utilizing heavy equipment versus labour to carry out roadworks.
- *Social benefits* – depicts the social benefits derived from using local resources, employment and local income generation, as opposed to the use foreign contractors and employees.
- *Quarrying options* – illustrates two options for quarry works using capital or labour-based technologies.
- *Trenching Options* – illustrates two options for carrying out trenching work; one equipment-based, the other labour intensive.

ITDG SRI LANKA (1996)
Routes –Rural transport needs
Video VHS-Pal, Colombo, 18 min.
Available from: ITDG Sri Lanka

This very professional video describes rural transport problems in Sri Lanka and some solutions through the provision of IMT. Firstly the film outlines the problems of bad rural accessibility with a number of impressive examples. The problems are illustrated with good pictures about footpaths, missing bridges, transport to hospitals, bad access to markets, loading of bicycles, transport with two-wheeled tractors and rudimentary transport services. The second part demonstrates the advantages of a cycle trailer and of an extended bicycle developed by ITDG Sri Lanka. Examples visualize the economic and social benefits generated by the use of these IMT: transport of goods and persons, use of IMT by traders and transport services. The video can be recommended as an introduction to rural transport problems.

STARKEY, PAUL (1995)
Portraying animal traction in South Africa: Empowering rural communities
DBSA and SANAT, Gauteng, South Africa
Available from: ATNESA

This booklet contains sixteen pages with 50 colour pictures about animal traction in Africa and Europe. The images can be used to depict various aspects of animal power: oxen, donkeys, mules, camels and horses carrying loads, pulling ploughs, carts and even cars. These pictures are part of a book by Starkey (1995): Animal Traction in South Africa, DBSA and SANAT, Gauteng, South Africa. The booklet is separately available from Paul Starkey (ATNESA).

AN ANNOTATED BIBLIOGRAPHY ON RURAL TRANSPORT

11. HOW TO OBTAIN THE DOCUMENTS

The documents can be obtained from different sources. Books with ISBN and ISSN numbers can be ordered through a bookshop. Publications from institutions like The World Bank, ILO, universities etc. can be retrieved from the documentation centre of these organizations or from special departments. Articles of journals and other unpublished documents can be ordered from the Information Services and Training Project ILO/ASIST in Nairobi, Kenya. Useful addresses and contact numbers are given below.

ILO/ASIST Nairobi
Advisory Support
Information Services and Training Project
P. O. Box 60598
Nairobi, Kenya
Tel: 254-2-560941, 560945
Fax: 254-2-566234
Email: iloasist@arcc.or.ke

ATNESA
Animal Traction Development
Oxgate, 64 Northcourt Avenue
Reading RG2 7HQ
United Kingdom
Tel: 44 118 987-2152
Fax: 44 118 931-4525
Email: P.H.Starkey@ reading.ac.uk

DTU
Development Technology Unit
Department of Engineering
University of Warwick
Coventry, CV4 7AL
United Kingdom
Tel: 44 1203 523 523 ext. 2339
Fax: 44 1203 418 922
Email: esceo@eng.warwick.ac.uk

FAO
Via delle Terme di Caracalla
00100 Rome
Italy
Fax: 39 6 5225-5749

GATE/GTZ
P.O. Box 5180
65 760 Eschborn
Germany
Tel.: 44 6196 790
Fax.: 44 6196 791-115

IFRTD
150 Southampton Row, 2nd floor
London WC1B 5AL
United Kingdom
Tel: 44 171 278-3670
Fax: 44 171 278-6880
Email: ifrtd@gn.apc.org

IHE
Westvest 7
P. O. Box 3015
2601 DA, Delft
The Netherlands
Tel: 31 15 215-1715
Fax: 31 15 212-2921
Email: ihe@ihe.nl

ILO/POLDEV Geneva
4, route des Morillons
1211Geneva 22
CH-1211 Switzerland
Fax: 41 22 799-6489, 798-8685

ILO/PUBL Geneva
4, route des Morillons
1211Genèva 22
CH-1211 Switzerland
Tel.: 41 22 799 61 11
Fax: 41 22 798 63 58

Intermediate Technology Publications
103-105 Southampton Row
London WC1B 4HH
United Kingdom
Tel: 44 171 436-9761
Fax: 44 171 436-2013
Email: itpubs@itpubs.org.uk

ITDG
Myson House, Railway Terrace
Rugby CV21 3HT
United Kingdom
Tel: +44 1788 560-631
Fax: +44 1788 540-270
Email: postmaster@itdg.org.uk

ITDG Sri Lanka
5, Lionel Edirisinghe Mawatha
Kirulapone, Colombo 5
Sri Lanka
Tel: 94 1 852-149
Fax: 94 1 856-188
Email: postmaster@itdg.lanka.net

IT Transport
Old Power Station
Ardington, Oxon OX12 8PH
United Kingdom
Tel: 44 1235 833-753
Fax: 44 1235 832-186
Email: ittran@rmple.co.uk

School of Development Studies
University of East Anglia
Norwich, NR4 7TJ
United Kingdom
Fax: 44 603 505-262

Silsoe College
Cranfield University
Bedfordshire MK45 4DT
United Kingdom

SKAT
Vadianstrasse 42
CH-9000 St. Gallen
Switzerland
Tel: 41 71 228-5454
Fax: 41 71 228-5455

TRL
Overseas Centre
Old Wokingham Road, Crowthorne
Berkshire RG45 6AU
United Kingdom
Tel: 44 1344 773 131
Fax: 44 1344 770-356

USAID
Public Enquiries
320 21st Street, N.W.
Washington, D.C. 20523-0016
Tel.: 1 202 647-1850
Fax: 1 202 647-8321

World Bank
Leita Jones
Task Team Assistant, SSATP
AFTTI - J11-292
1818 H Street, N.W.
Washington, D.C. 20433
U. S. A.
Tel.: 1 202 473-5030
Fax: 1 202 473-8326
Email: LJones2@Worldbank.org

World Bank Bookstore
P. O. Box 7247-8619
Philadelphia, PA 19170
U. S. A.
Tel.: 1 703 661-1501
Fax: 1 703 661-1580